Advanced Structured Materials

Volume 17

Series Editors
Andreas Öchsner
Lucas F. M. da Silva
Holm Altenbach

For further volumes:
http://www.springer.com/series/8611

Mohd Nasir Tamin
Editor

Damage and Fracture of Composite Materials and Structures

Springer

Mohd Nasir Tamin
Faculty of Mechanical Engineering
Universiti Teknologi Malaysia
UTM Skudai
81310 Johor
Malaysia
e-mail: taminmn@gmail.com

ISSN 1869-8433
ISBN 978-3-642-23658-7
DOI 10.1007/978-3-642-23659-4
Springer Heidelberg Dordrecht London New York

e-ISSN 1869-8441
e-ISBN 978-3-642-23659-4

Library of Congress Control Number: 2011941716

© Springer-Verlag Berlin Heidelberg 2012
This work is subject to copyright. All rights are reserved, whether the whole or part of the material is concerned, specifically the rights of translation, reprinting, reuse of illustrations, recitation, broadcasting, reproduction on microfilm or in any other way, and storage in data banks. Duplication of this publication or parts thereof is permitted only under the provisions of the German Copyright Law of September 9, 1965, in its current version, and permission for use must always be obtained from Springer. Violations are liable to prosecution under the German Copyright Law.
The use of general descriptive names, registered names, trademarks, etc. in this publication does not imply, even in the absence of a specific statement, that such names are exempt from the relevant protective laws and regulations and therefore free for general use.

Cover design: WMXDesign GmbH, Heidelberg

Printed on acid-free paper

Springer is part of Springer Science+Business Media (www.springer.com)

Preface

This book compiles recent advances in research and development in the area of damage and fracture of composite materials and structures. Majority of these chapters have been presented and discussed at the Special Session on Damage and Fracture of Composite Materials and Structures of the 4th International Conference on Advanced Computational Engineering and Experimenting (ACE-X), held during 8–10 July, 2010 in Paris, France. Invited chapters on selected topics are included. The scope of the book covers experimental, analytical and computational aspects of the mechanics of fibrous composites. The chapters report on updates related to testing of composite samples for determination of mechanical properties, discussion on the observed failure mechanisms and numerical modeling of deformation response of the composites to loading. Behavior of fibrous composites, sandwich structures and fiber-metal laminates to impact and blast loading are discussed. Response of composite structures under both static and dynamic loading is described.

Each chapter is written as a self-contained report on recent research works of the respective authors. However, many issues related to damage and fracture of the composite are common. These include dominant fracture mechanisms, damage-based models and simulation approaches. The book is suitable not only for research, but also for advanced reference and applications.

We acknowledge contributions from authors, many of whom are principal researchers of their projects on characterization of composite materials and structures. Thank you to Hassan Osman Ali for assistance in preparing this monograph.

Johor, Malaysia, 2011   M. N. Tamin

Contents

Introduction 1
M. N. Tamin

**Mode-I Crack Control by SMA Fiber with
a Special Configuration** 5
M. Jin and S. H. Chen

**Micromechanical Analysis of Mode I Crack Growth
in Carbon Fibre Reinforced Polymers** 17
Daniel Trias and Pere Maimí

**Modeling of Spalling Effect on Toughening in Fiber
Reinforced Composites** 27
C. Wang, N. F. Piaceski and K. M. Soares

**Evolution Characteristics of Delamination Damage
in CFRP Composites Under Transverse Loading** 45
S. S. R. Koloor, A. Abdul-Latif, X. J. Gong and M. N. Tamin

**Indentation of Sandwich Beams with Functionally Graded Skins
and Transversely Flexible Core** 61
Y. Mohammadi and S. M. R. Khalili

**Micromechanical Fibre-Recruitment Model of Liquid Crystalline
Polymer Reinforcing Polycarbonate Composites** 85
K. L. Goh and L. P. Tan

Moisture Absorption Effects on the Resistance to Interlaminar Fracture of Woven Glass/Epoxy Composite Laminates 107
X. J. Gong, K. J. Wong and M. N. Tamin

The Study of Response of High Performance Fiber-Reinforced Composites to Impact Loading............................. 129
O. Saligheh, R. Eslami Farsani, R. Khajavi and M. Forouharshad

Dynamic Fracture Toughness of Composite Materials 143
C. Rubio-González, J. Wang, J. Martinez and H. Kaur

Impact Study on Aircraft Type Laminate Composite Plate; Experimental, Failure Criteria and Element Model Review 157
Y. Aminanda

The Blast Response of Sandwich Structures..................... 189
M. Yazid Yahya, W. J. Cantwell, G. S. Langdon and G. N. Nurick

The High Velocity Impact Response of Self-Reinforced Polypropylene Fibre Metal Laminates 219
M. R. Abdullah and W. J. Cantwell

Index 241

Introduction

M. N. Tamin

Abstract Fiber-reinforced composites cover wide range of composite materials and structures including laminated panels of unidirectional plies, sandwich structures with fiber-reinforced composite skins and flexible core, and fiber-metal laminates. A typical advanced composite structure will likely experience quasi-static and dynamic loading. Under such damage-induced operating load conditions, reliability of the structure largely depends on the continual process of damage initiation and subsequent damage propagation to catastrophic fracture. The development of validated damage and failure models and availability of vast computing power would enable establishment of a validated framework for addressing structural reliability and safety issues of these composite structures.

Keywords Fiber-reinforced composites · Damage-based models · Continuum damage mechanics · Damage mechanisms · Fatigue

Fiber-reinforced composites cover wide range of composite materials and structures such as laminated panels of unidirectional plies, sandwich structures with fiber-reinforced composite skins and flexible core, and fiber-metal laminates (FML). Although glass and carbon fibers is relatively common, polymer, metal and ceramic fibers are being used as reinforcement phase in specific applications. Fiber-reinforced polymer (FRP) matrix composites offer an alternate replacement material to high-strength metals. In the transportation sector, the use of light-weight carbon fiber-reinforced polymer (CFRP) composites in airframe, automotive and railcar structures leads to improved (lower) fuel consumption and/or increased payload. The growth in use of CFRP composites from 15% in 1990

M. N. Tamin (✉)
Faculty of Mechanical Engineering, Center for Composites,
Universiti Teknologi Malaysia (UTM), 81310 Skudai, Johor, Malaysia
e-mail: taminmn@fkm.utm.my

(A320) to over 50% in 2010 (A380) for airframe structures and high lift components is demonstrated [1]. The increasing use of CFRP composites is derived from its high strength-to-weight ratio, high specific modulus and flexible design through sequencing of the pre-impregnated laminates for tailored strength and stiffness properties in particular loading direction. The composite's resistance to environmental corrosion is an added advantage over their metallic counterpart. This intrinsic property is appealing to applications in construction industry for bridges and wastewater and chemical storage tanks. Other property such as the composites transparency to radar is valuable for stealth applications.

In numerous applications the composite structure will be subjected to both static and transient load throughout the design lives. As such, it is essential to identify dominant damage mechanisms and quantify failure process of the composite materials. This calls for evaluation of the materials resistance to quasi-static, impact and fatigue loading along with establishment of mechanical damage evolution. Reliability of the composite structure under such damage-induced operating load conditions depends on the continual process of damage initiation and subsequent damage propagation to catastrophic fracture. Such reliability data is indispensable during both initial material selection and detailed engineering design stage of a composite structure. Simple tests of the laminated composite coupons under tension, compression, shear and flexural loading are useful to measure intrinsic properties of the composites [2–4]. Fracture tests on pre-cracked samples are often used to establish fracture toughness and critical energy release rates of the composites in Mode I and II loadings [5]. However, specially-designed specimen, jigs and fixtures, and environmental chamber are required to establish mixed-mode loading effects [6], mean stress effects on fatigue life and moisture conditioning effects in accelerated testing [7]. The FRP composites ability to be fabricated into a wide range of laminated configurations can create costly materials evaluation programs. The magnitude of the test program could be reduced by implementing reliable life prediction models based on the properties and macro-lay-ups of the laminates.

Failure process in FRP composites is a complex phenomenon. The different failure modes such as matrix yielding, matrix cracking, fiber/matrix interface de-bonding, fiber pull-out, fiber fracture and inter-ply delamination could occur in isolation or synergize simultaneously in FRP composites. An example of the various failure modes for unidirectional SiC fiber-reinforced titanium metal matrix composite is shown in Fig. 1. Additional fracture mechanisms observed following high-velocity impact and blast load are sheared, deformed and melted fibers, ductile tearing, thinning and sheared fracture of the metal in FMLs. Interface delamination in CFRP composites is of particular interest in view of the relatively weak ply-to-ply interface strength and the large interface shear stresses developed during loading. Interface delamination failure leads to significant loss of the load-carrying capacity and occurs in the absence of any visible damage of the CFRP composite [8]. However, bridging crack growth observed in continuous fiber-reinforced composites contributes to fracture toughness of the material through load shedding mechanisms [9]. Such critical delamination failure and bridging crack growth process could be accounted for in the design phase with the aid of

Fig. 1 Bridging crack growth mechanism in SiC fiber-reinforced Ti-metal matrix composite illustrating fiber pull-out, fiber fracture, fiber/matrix interface debonding and matrix cracking. Fatigue crack growth test at 500 C, 10 Hz [9]

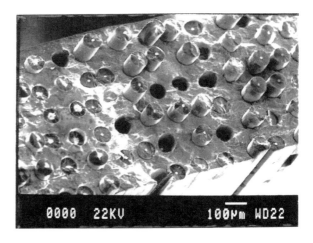

numerical modeling and validation testing of FRP composite specimens. In this respect, several materials models have been proposed and examined for predicting the composite strength with the different types of constituent defects. These include stress-based failure models [10–12] and strain-based criteria [13, 14] for matrix cracking and fiber fracture. Complete treatment of the mechanics of composite materials and the various failure models is best described in textbooks and handbooks [15–17]. The progression of each type of defects under fatigue load cycles dictates the useful life of the part. In this respect, cohesive zone model has been formulated for predicting ply interface delamination damage initiation and subsequent propagation under monotonic loading condition [18, 19]. The damage-based model employing virtual crack closure concept is introduced to account for load reversals in fatigue failure of the laminates [20, 21].

The simultaneous occurrence of multiple failure modes in a composite part under loading could not be represented by any of these models in isolation. To this end a unified approach based on continuum damage mechanics concept is much needed to account for the synergy effects of multiple failure mechanisms in FRP composites. Implementation of the newly-developed damage-based models into commercial finite element codes (e.g. [22]) and availability of vast computing power would yield a validated framework for addressing structural reliability and safety issues of FRP composite structures.

References

1. Baker, A.A., et al., (eds.).: Composite Materials for Aircraft Structures, 2nd edn. AIAA Education Series, USA (2004)
2. Standard Test Method for Tensile Properties of Polymer Matrix Composite Materials. ASTM Standard D3039/D3039 M-95a

3. Standard Test Method for Compressive Properties of Unidirectional or Crossply Fiber-resin Composites. ASTM Standard D3410/3410 M-95
4. Standard Test Method for Shear Properties of Composite Materials by the V-Notched Beam Method. ASTM Standard D5379/D5379 M-93
5. Mode I Interlaminar Fracture Toughness of Unidirectional Fiber-Reinforced Polymer Matrix Composites, ASTM Standard D5528 (1994)
6. Benzeggagh, M.L., Kenane, M.: Measurement of Mixed-Mode Delamination Fracture Toughness of Unidirectional Glass/Epoxy Composites with Mixed-Mode Bending Apparatus. Compos. Sci. Technol. **49**, 439–449 (1996)
7. Ciriscioli, P.R., Lee, W.I., Peterson, D.G.: Accelerated Environmental Testing of Composites. J. Compos. Mater. **21**, 225–242 (1987)
8. Koloor, S.S.R., Abdul-Latif, A., Tamin, M.N.: Mechanics of Composite Delamination under Flexural Loading. Key Eng. Mater. **462–463**, 726–731 (2011)
9. Tamin, M.N., Ghonem, H.: Fatigue Damage Mechanisms of Bridging Fibers in Titanium Metal Matrix Composites. J. Eng. Mater. Technol. Trans. ASME **122**, 370–375 (2000)
10. Sun, C.T., Tao, J.X.: Prediction Failure Envelopes and Stress/Strain Behavior of Composite Laminates. Compos. Sci. Technol. **58**, 1113–1125 (1998)
11. Hashin, Z., Rotem, A.: A Fatigue Failure Criterion for Fiber Reinforced Materials. J. Compos. Mater. **7**, 448–464 (1973)
12. Tsai, S.W., Wu, E.M.: A General Theory of Strength for Anisotropic Materials. J. Compos. Mater. **5**, 58–80 (1971)
13. Hart-Smith, L.J.: Prediction of the Original and Truncated Maximum Strain-Strain Failure Models for Certain Fibrous Composite Laminates. Compos. Sci. Technol. **58**, 1151–1178 (1998)
14. Eckold, G.G.: Failure Criteria for Use in design Environment. Compos. Sci. Technol. **58**, 1095–1105 (1998)
15. Chawla, K.K.: Composite Materials Science and Engineering. Springer, New York (1998)
16. Baker, A., Dutton, S., Kelly, D. (eds.).: Composite Materials for Aircraft Structures, 2nd edn. AIAA Inc., (2004)
17. Mallick, P.K. (ed.).: Composite Materials Handbook. Marcel Dekker, New York (1997)
18. Davila, C.G., Camanho, P.P., Moura, M.F.: Mixed-Mode Decohesion Elements for Analyses of Progressive Delamination. In: Proceedings of 42nd AIAA/ASME/ASCE/AHS/ASC Structures, Structural Dynamics and Materials Conference, Seattle, WA, 16–19 April (2001)
19. Turon, A., Camanho, P.P., Costa, J., Davila, C.G.: An Interface Damage Model for the Simulation of Delamination Under Variable-Mode Ratio in Composite Materials. NASA/TM-2004-213277. VA, USA (2004)
20. Krueger, R.: The Virtual Crack Closure Technique: History, Approach and Application. NASA/CR-2002-211628. NTIS, VA, USA (2002)
21. Xie, D., Biggers Jr., S.B.: Progressive Crack growth Analysis using Interface Element based on Virtual Crack Closure Technique. Finite Elem. Anal. Desi. **42**, 977–984 (2006)
22. Abaqus ver 6.9, Simulia Inc.

Mode-I Crack Control by SMA Fiber with a Special Configuration

M. Jin and S. H. Chen

Abstract Crack propagation in solid members is an important reason for structure failure. In recent years, many research interests are focused on intelligent control of crack propagation. With the rise in temperature, contraction of prestrained SMA fiber embedded in matrix makes retardation of crack propagation possible. However, with the rise in temperature, separation of SMA fiber from matrix is inevitable. This kind of separation weakens effect of SMA fiber on crack tip. To overcome de-bonding of SMA fiber from matrix, a knot is made on the fiber in this paper. By shape memory effect with the rise in temperature, the knotted SMA fiber generates a couple of recovery forces acting on the matrix at the two knots. This couple of recovery forces may restrain opening of the mode-I crack. Based on Tanaka constitutive law on SMA fiber and complex stress function near an elliptic hole under a point load, a theoretical model on mode-I crack is proposed. An analytical expression of relation between stress intensity factor (SIF) of mode-I crack closure and temperature is got. Simulation results show that stress intensity factor of mode-I crack closure decreases obviously with the rise in temperature higher than the austenite start temperature of SMA fiber, and that there is an optimal position for SMA fiber to restrain crack opening, which is behind the crack tip. Therefore the theoretical model supports that prestrained SMA fiber with knots in martensite can be used to control mode-I crack opening effectively because de-bonding between fiber and matrix is eliminated. Specimen of epoxy resin embedded with knotted SMA fiber can be made in experiment and is useful to an

M. Jin (✉) · S. H. Chen
The State Key Laboratory of Nonlinear Mechanics (LNM),
Beijing 100190, People's Republic of China
e-mail: mjin@bjtu.edu.cn

M. Jin
Department of Mechanics, School of Civil Engineering, Beijing Jiaotong University,
Beijing 100044, People's Republic of China

analytical study. However, in practical point of view, SMA fiber should be embedded in engineering structure material such as steel, aluminum, etc. The embedding process in these matrix materials should be studied systematically in the future.

Keywords Crack control · SMA fiber · Actuator · Complex stress function

1 Introduction

Crack propagation in solid members is an important reason for structure failure. Many researchers are interested in methods of crack control. With the rise in temperature, contraction of prestrained shape memory alloy (SMA) fiber embedding in matrix makes retardation of crack propagation possible. Rogers et al. [1] proposed that embedded prestrained SMA fiber near crack tip may reduce the stress by contraction effect of SMA under increase of temperature. Photoelastic pictures and finite element analysis show that stress intensity factor (SIF) of a specimen embedded with prestrained SMA fiber decreased as temperature increases [2–4]. To overcome separation of SMA fiber from matrix, surface of SMA fiber is treated chemically [2]. Shimamoto et al. [5–8] and Umezaki et al. [9] indicated that stress at the tip of mode-I crack in epoxy matrix with SMA fiber can be reduced by heating SMA fiber. Quality change in matrix near SMA fiber is observed if the temperature is too high. This kind of quality change damages the contraction effect of SMA fiber on crack closure in matrix. Araki et al. [10, 11] performed a micromechanical model on crack closure considering crack bridging fibers and compared their numerical results with experiment. Experiments in these researches give a possible way in practical use to control crack propagation in a specimen.

However, with the rise in temperature, separation of SMA fiber from matrix is inevitable. This kind of separation weakens effect of SMA fiber on crack tip. In other words, effect of those applications is limited by the interfacial debonding. Umezaki [12] proposed a type of spiral SMA wire to prevent separation of SMA wire from matrix. In order to avoid the interface de-cohesion or sliding, Gabry et al. [13] used different surface treatments on SMA wires. Pullout tests and scanning electronic microscopy observations [14] indicated that increasing the number of turns of the twisted NiTi wires strengthens bonding strength between fiber and matrix. Zhou et al. [15–17] studied SMA fiber debonding in matrix. Wang et al. [18] proposed a theoretical model for SMA pullout from an elastic matrix. Hu et al. [19] studied interface failure between SMA fiber and matrix by shear lag model. Considering debonding at the interface between SMA wires and a polymer matrix, Wang [20] estimated the minimum volume fraction of SMA wires required to produce stress for the active crack closure. By finite element simulation, Burton et al. [21] indicated that reverse transformation of SMA wires during heating causes a closure force across the crack.

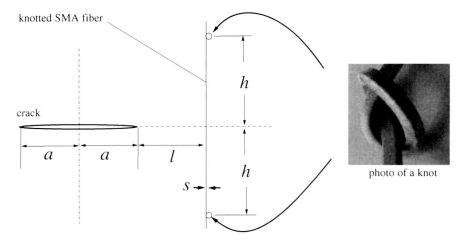

Fig. 1 A knotted prestrained SMA fiber perpendicular to crack surface

To improve effect of SMA fiber on matrix, small knots on SMA fiber are made by authors. Figure 1 shows configuration of a prestrained SMA fiber with knots at each end. With the rise in temperature, the SMA fiber contracts and the two knots on the fiber locking tightly on matrix by recovery stress. This recovery force may reduce the stress intensity factor at the crack tip and make the crack closure. In this process, the fiber cannot be debonded from matrix totally. Therefore, as an actuator in practical use, knotted SMA fiber can be embedded in matrix to control crack propagation. Relation of stress intensity factor with temperature is an important problem in crack control. In this paper, an approximate model on stress intensity factor of mode-I crack closure in matrix is proposed on a specimen with embedded SMA fiber with knots.

2 Relation of Recovery Force on Matrix with Temperature

See Fig. 1. Considering a specimen with a mode-I crack in matrix embedded with a prestrained SMA fiber with knots in martensite. SMA fiber is perpendicular to crack surface. Plane strain problem is considered.

$2a$ and $2h$ denote length of crack and length between the two knots, respectively. l denotes distance from the right crack tip to the fiber. With the rise in temperature, the knotted SMA fiber contracts due to phase transformation from martensite to austenite. The SMA fiber generates a couple of recovery forces $-P$ at upper knot Z_0 and P at lower knot $\bar{Z}_0$ acting on matrix where

$$Z_0 = a + l + ih, \qquad (1)$$

Fig. 2 A couple of recovery forces, P acting on matrix

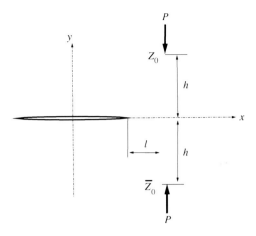

$\bar{Z}_0$ denotes conjugate of Z_0; and i denotes imaginary unit. See Fig. 2. Because bonding on interface between fiber and matrix is much weaker than locking between knot and matrix, interfacial shear stress is neglected in this paper. Therefore, the knotted SMA fiber is in simple tension by axial load, P. We have

$$P = s\sigma_f, \qquad (2)$$

where σ_f and s denote recovery stress and area of the SMA fiber, respectively. In phase transformation, strain of SMA fiber is about several percent which is one order higher than strain of epoxy in elastic range. So we neglect strain of matrix when we consider strain of SMA fiber. Because each knot is locked in the matrix as the fiber contracts, increase of normal strain of SMA fiber embedding in the matrix is approximately zero, i.e.

$$\delta\varepsilon_f = 0, \qquad (3)$$

where ε_f denotes normal strain of SMA fiber in phase transformation. Based on the Tanaka one-dimensional constitutive law on SMA [22], we have

$$\delta\sigma_f = E_f\,\delta\varepsilon_f + \Theta\,\delta T + \Omega\,\delta\xi, \qquad (4)$$

where E_f denotes Young's modulus; Θ denotes thermoelastic modulus; and Ω denotes phase transformation modulus. If ε_L denotes maximum residual strain, we have $\Omega = -E_f\varepsilon_L$ in SMA fiber. $\delta\sigma_f$, δT and $\delta\xi$ denote increase of normal stress, rise in temperature and increase of martensitic volume fraction of SMA fiber. Martensitic volume fraction, ξ is a function of stress, σ_f and temperature, T, i.e.,

$$\xi = \xi(\sigma_f, T). \qquad (5)$$

Substituting Eqs. 3 to 5 into 4, we have increase of recovery stress

Fig. 3 Recovery stress, σ_f versus temperature, T

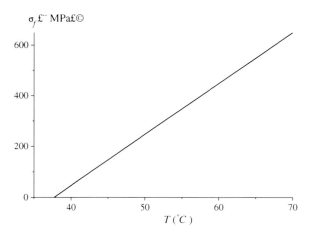

$$\delta\sigma_f = \frac{\Theta + \Omega \frac{\partial \xi}{\partial T}}{1 - \Omega \frac{\partial \xi}{\partial \sigma_f}} \delta T. \tag{6}$$

In phase transformation from martensite to austenite with rise in temperature, function (5) can be approximated by an exponential function [22] with respect to temperature, T, i.e.,

$$\xi = \exp\left[a_A(A_s - T) + b_A\sigma_f\right], \tag{7}$$

where $a_A = \frac{-2\ln 10}{A_s - A_f}$; $b_A = \frac{a_A}{C_A}$; A_s and A_f are austenite start temperature and austenite finish temperature of SMA fiber, respectively; C_A denotes influence coefficient of stress on transition temperature.

Substituting Eq. 7 into 6, we have

$$\delta\sigma_f = \frac{\Theta - a_A\Omega\exp\left[a_A(A_s - T) + b_A\sigma_f\right]}{1 - b_A\Omega\exp\left[a_A(A_s - T) + b_A\sigma_f\right]} \delta T. \tag{8}$$

Before heating the SMA fiber, stress in the SMA fiber is zero as temperature, $T \leq A_s$. The initial value of σ_f in Eq. 8 is $\sigma_f|_{T=A_s} = 0$. Generally, we can solve initial value problem (8) numerically by the Euler iteration. However, in Appendix A, we have an analytical solution of initial value problem (8) on martensitic volume fraction, ξ

$$\begin{cases} T = \dfrac{\ln\xi - b_A\Omega(\xi - 1)}{b_A\Theta - a_A} + A_s \\ \sigma_f = \dfrac{\ln\xi - a_A(A_s - T)}{b_A} \end{cases} \tag{9}$$

where ξ decreases at 1. For a given ξ, we can get T and σ_f from Eq. 9. Substituting the value of σ_f into Eq. 2, we have relation of point load, P with temperature, T. Figure 3 shows curve of recovery stress, σ_f versus temperature, T.

Table 1 Material Properties of NiTi Fiber [16, 21]

E_f	ε_L	A_s	A_f	Θ	C_A
37.4 GPa	0.067	37.7°C	47.2°C	9.1 MPa/°C	20.3 MPa/°C

3 Stress Intensity Factor of Mode-I Crack Closure Under a Couple of Point Loads

Complex stress function of an elliptic hole in plane strain under a point load is given by [23]. The elliptic hole degenerates into a crack in Fig. 1 as perpendicular axis of elliptic hole tends to zero. Thus we can get complex stress function of the crack under a point load. Based on this complex stress function, we can get stress intensity factor of mode-I crack closure under a couple of point loads P in Fig. 2. The formulation can be seen in Appendix B. Considering Eq. 2, we have

$$\frac{\sqrt{a}}{s}K_I = \frac{\sigma_f}{\sqrt{\pi}}\left\{2\mathrm{Im}\left(\frac{1}{\zeta_0 - 1}\right) + (1+v')\mathrm{Im}\left(\zeta_0 + \frac{1}{\zeta_0}\right)\mathrm{Re}\left[\frac{\zeta_0^2}{(\zeta_0+1)(\zeta_0-1)^3}\right]\right\}, \quad (10)$$

where K_I is the stress intensity factor of mode-I crack closure, ζ_0 is given by $Z_0 = \frac{a}{2}\left(\zeta_0 + \frac{1}{\zeta_0}\right)$; $v' = \frac{v}{1-v}$ where v denotes Poisson's ratio of matrix; Re() and Im() denote real component and imaginary component of (), respectively. Recovery stress, σ_f can be calculated numerically from Eq. 8 under a temperature, T or analytically from Eq. 9. Substituting recovery stress, σ_f into Eq. 10, we have the stress intensity factor, K_I of mode-I crack closure. Then we have relation of the stress intensity factor, K_I of mode-I crack closure with temperature, T.

4 Numerical Results

Material properties of SMA fiber are listed in Table 1, where E_f denotes Young's modulus; ε_L denotes maximum residual strain; A_s denotes austenite start temperature; A_f denotes austenite finish temperature; Θ denotes thermoelastic modulus; C_A denotes influence coefficient of stress on transition temperature. Poisson's ratio of matrix is $v = 0.3$.

Figure 3 shows curve of recovery stress, σ_f versus temperature, T. It is shown that recovery stress, σ_f reaches about 600 MPa as temperature rises to 70°C. Figure 4 shows martensitic volume fraction, ξ as a function with respect to temperature, T in phase transformation in Figs. 5, 6, 7, 8 and 9. It is shown that martensitic volume fraction, ξ decreases as the temperature rises from 40 to 70°C.

Fig. 4 Martensitic volume fraction, ξ versus temperature, T

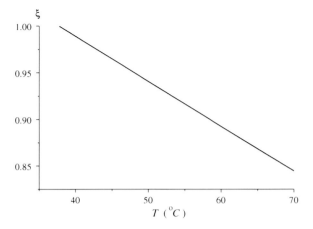

Fig. 5 The closure stress intensity factor, K_I versus temperature, T for different fiber position as $h/a = 1.0$

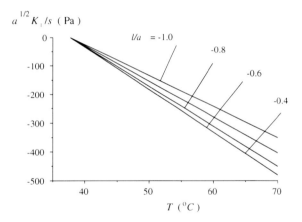

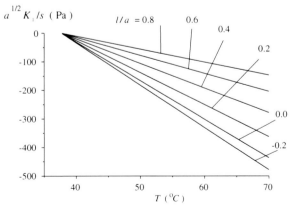

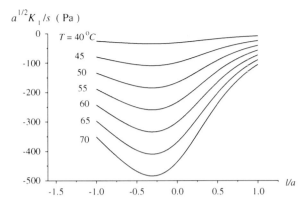

Fig. 6 The closure stress intensity factor, K_I versus fiber position, l/a at different temperature as $h/a = 1.0$

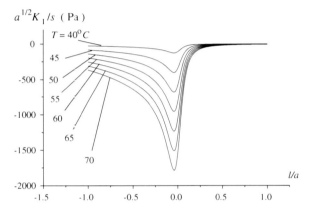

Fig. 7 The closure stress intensity factor, K_I versus fiber position, l/a at different temperature as $h/a = 0.1$

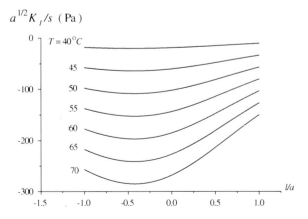

Fig. 8 The closure stress intensity factor, K_I versus fiber position l/a at different temperature as $h/a = 2.0$

As crack length is the same as fiber length, i.e. $\frac{h}{a} = 1.0$, Fig. 5 shows the stress intensity factor of mode-I crack closure as function with respect to temperature for different fiber position. It is shown that stress intensity factor decreases almost

Fig. 9 The closure stress intensity factor, K_I versus fiber position l/a at different fiber length

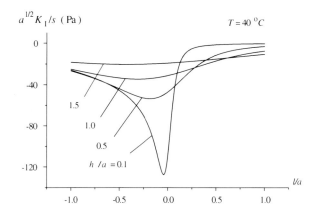

linearly with rise in temperature. It is shown that stress intensity factor may take the minimum value at the same temperature between $\frac{l}{a} = -0.2$ and -0.4. Figure 6 shows closure stress intensity factor as a function with respect to fiber position at different temperature. At the same temperature, the nearer the fiber locates the crack tip, the little the stress intensity factor is. Effect of crack closure of the fiber behind crack tip is more obvious than that of the fiber in front of crack tip. It is also shown that there is an optimal position for fiber where the stress intensity factor takes the minimum value. For $\frac{h}{a} = 1.0$, this optimal position for effect of crack closure is approximately at $\frac{l}{a} = -0.3$ behind the crack tip. Figures 7 and 8 show the stress intensity factor of mode-I crack closure as a function with respect to fiber position at different temperature for $\frac{h}{a} = 0.1$ and 2.0, respectively. Approximately $\frac{l}{a} = -0.04$ and -0.45 are the optimal position for effect of crack closure for $\frac{h}{a} = 0.1$ and 2.0, respectively.

Figure 9 shows the closure stress intensity factor as a function with respect to fiber position $\frac{l}{a}$ for different fiber length $\frac{h}{a}$ at temperature 40°C. Near the crack tip, it is shown that, the more short distance of the two knots is, the more obvious the effect of crack closure is. In other words, the nearer the knot locates the crack tip, the more obvious effect of crack closure is.

5 Conclusions

Based on the Tanaka one-dimensional constitutive law on SMA, an approximate model on stress intensity factor of mode-I crack closure by prestrained knotted SMA fiber in martensite is proposed. Relation of the stress intensity factor of mode-I crack closure with temperature is given in Eqs. 8 and 9. Numerical results show that the stress intensity factor of mode-I crack closure increases as the prestrained knotted SMA fiber in martensite is heated. Optimal position of the knotted SMA fiber for crack closure is behind the crack tip. As an actuator in matrix, prestrained knotted SMA fiber in martensite can be used to control mode-I crack opening.

Acknowledgments Authors gratefully acknowledge support from Opening Fund of State Key Laboratory of Nonlinear Mechanics (LNM) of Chinese Academy of Sciences, P. R. China, through grant number LNM201003 and the Science Foundation of Beijing Jiaotong University through grant number PD-225.

Appendix A

From derivative of Eq. 7 with respect to temperature, T, we have

$$\frac{\delta \sigma_f}{\delta T} = \frac{1}{b_A}\left(\frac{\delta \xi}{\xi \delta T} + a_A\right). \tag{A1}$$

Substituting Eqs. 7 and (A1) into 8, we have

$$\left(\frac{1}{\xi} - b_A \Omega\right)\delta \xi = (b_A \Theta - a_A)\delta T. \tag{A2}$$

Indefinite integral of Eq. A2 is

$$T = \frac{\ln \xi - b_A \Omega \xi}{b_A \Theta - a_A} + c. \tag{A3}$$

where c is a constant. Substituting $\sigma_f|_{T=A_s} = 0$ into Eqs. 7 and A3, we have

$$c = \frac{b_A \Omega}{b_A \Theta - a_A} + A_s. \tag{A4}$$

Substituting Eq. A4 into A3, we have the first equation in Eq. 9. The second equation in Eq. 9 can be got easily from Eq. 7.

Appendix B

Complex stress functions $\Phi(\zeta)$ and $\chi(\zeta)$ of an elliptic hole in plane strain under a point load P at Z_0 in Fig. 2. are given in Eqs. 4 and 5 in article of [23], respectively. As perpendicular axis of the elliptic hole tends to zero, we have derivatives of $\Phi(\zeta)$ and $\chi(\zeta)$ with respect to ζ

$$\frac{d\Phi(\zeta)}{d\zeta} = \frac{iPa^2}{8\pi}\left\{(3-v')\left(\frac{1}{\zeta - \frac{1}{\overline{\zeta_0}}} - \frac{1}{\zeta}\right) + (1+v')\left[\frac{1}{\zeta - \zeta_0} + \frac{\left(\overline{\zeta_0} + \frac{1}{\overline{\zeta_0}}\right) - \left(\zeta_0 + \frac{1}{\zeta_0}\right)}{\left(\overline{\zeta_0}^2 - 1\right)\left(\zeta - \frac{1}{\overline{\zeta_0}}\right)^2}\right]\right\} \tag{B1}$$

and

$$\frac{d\chi(\zeta)}{d\zeta} = \frac{iPa^2}{8\pi}\left\{\frac{3-v'}{\zeta-\zeta_0} + (1+v')\left[\frac{1}{\zeta-\frac{1}{\bar{\zeta}_0}} - \frac{1}{\zeta} + \frac{\left(\bar{\zeta}_0+\frac{1}{\bar{\zeta}_0}\right) - \left(\zeta_0+\frac{1}{\zeta_0}\right)}{(\zeta-\zeta_0)^2\left(1-\frac{1}{\zeta_0^2}\right)}\right]\right\}, \quad (B2)$$

where i is imaginary unit; ζ and ζ_0 are given by $Z = \frac{a}{2}\left(\zeta+\frac{1}{\zeta}\right)$ and $Z_0 = \frac{a}{2}\left(\zeta_0+\frac{1}{\zeta_0}\right)$, respectively; $2a$ is the crack length; $v' = \frac{v}{1-v}$ where v denotes Poisson ratio of matrix; and $\bar{\zeta}_0$ denotes conjugate of ζ_0. Perpendicular normal stress σ_y at Z in matrix under point load P at Z_0 in Fig. 2 is

$$\sigma_y = \frac{2\zeta^2}{a(\zeta^2-1)}\text{Re}\left(\frac{d\Phi}{d\zeta} + \frac{d\chi}{d\zeta}\right), \quad (B3)$$

where Re() denotes real component of ().

The closure stress intensity factor under point load P at Z_0 in Fig. 2 is

$$K_{I+} = \lim_{x \to 0} \sqrt{2\pi x}\sigma_y\big|_{y=0}. \quad (B4)$$

Substituting Eqs. B1, B2 and B3 into B4, we have expression of closure stress intensity factor K_{I+}. Similarly, we have expression of closure stress intensity factor K_{I-} under point load $-P$ at $\bar{Z}_0$ in Fig. 2. Expression of the closure stress intensity factor K_I under point load P at Z_0 and $-P$ at $\bar{Z}_0$ together in Fig. 2 can be obtained by

$$K_I = K_{I+} + K_{I-} \quad (B5)$$

Substituting Eq. 2 into Eq. B5, we have expression of K_I in Eq. 10.

References

1. Rogers, C.A., Liang, C., Lee, S.: Active damage control of hybrid material systems using induced strain actuator. In: Proceedings of 32nd Conference on Structures, Structural Dynamics and Materials, p. 1190 (1991)
2. Du, Y.L., Nie, J.X., Zhao, C.Z.: A new method for crack-detecting and active control. ACTA Aeronautica Et Astronautica Sinica **14**(7), A337–A341 (in Chinese) (1993)
3. Zhang, X.R., Nie, J.X., Du, Y.L.: Study on SMA intelligent structures for crack control. J. of Aerospace Power **13**(4), 357–361 (in Chinese) (1998)
4. Zhang, X.R., Nie, J.X, Du, Y.L.: Finite element analysis of smart structures for active crack control. ACTA Materiae Compositae Sinica **16**(2), 147–151 (in Chinese) (1999)
5. Shimamoto A., Taya M.: Reduction in K_I by the shape-memory effect in a TiNi shape-memory fiber-reinforced epoxy matrix composite. Trans. Japan Soc. Mech. Eng. A **63** (605), 26–31 (1997)

6. Shimamoto, A., Azakami, T., Oguchi, T.: Reduction of K_I and K_{II} by the shape-memory effect in a TiNi shape-memory fiber-reinforced epoxy matrix composite. Exp. Mech. **43** (1), 77–82 (2003)
7. Shimamoto, A., Okawara, H., Nogata, F.: Enhancement of mechanical strength by shape memory effect in TiNi fiber-reinforced composites. Eng. Fract. Mech. **71**, 737–746 (2004)
8. Shimamoto, A., Zhao, H., Azakami, T.: Active control for stress intensity of crack-tips under mixed mode by shape memory TiNi fiber epoxy composites. Smart Mater. Struct. **16**, N13–N21 (2007)
9. Umezaki, E., Kawahara, E., Watanabe, H.: Evaluation of crack closure in intelligent structure member with embedded SMA. JSME Int. J., Ser. C **41** (3), 470–475 (1998)
10. Araki, S., Ono, H., Saito, K.: Micromechanical analysis of crack closure mechanism for intelligent material containing TiNi fibers—1st report, modeling of crack closure mechanism and analysis of stress intensity factor. JSME Int. J. Ser. A-Solid Mech. Mater. Eng. **45**(2), 208–216 (2002)
11. Araki, S., Ono, H., Saito, K.: Micromechanical analysis of crack closure mechanism for intelligent material containing TiNi fibers—2nd report, numerical calculation of stress intensity factor in the process of shape memory shrinkage of TiNi fibers. JSME Int. J. Ser. A—Solid Mech. Mater. Eng. **45**(3), 356–362 (2002)
12. Umezaki, E.: Improvement in separation of SMA from matrix in SMA embedded smart structure. Mater. Sci. Eng. **A285**, 363–369 (2000)
13. Gabry, B., Thieb, F., Lexcellent, C.: Topographic study of shape memory alloy wires used as actuators in smart materials. J. Intell. Mater. Syst. Struct. **11**, 592–603 (2000)
14. Lau, K.T., Tam, W.Y., Meng, X.L., Zhou, L.M.: Morphological study on twisted NiTi wires for smart composite systems. Mater. Lett. **57**, 364–368 (2002)
15. Lau, K.T., Chan, W.L., Shi, S.Q., Zhou, L.M.: Debond induced by strain recovery of an embedded NiTi wire at a NiTi/epoxy interface: micro-scale observation. Mater. Des.**23**, 265–270 (2002)
16. Poon, C.K., Lau, K.T., Zhou, L.M.: Design of pull-out stresses for prestrained SMA wire/polymer hybrid composites. Compos. Part B **36**, 25–31 (2005)
17. Poon, C.K., Zhou, L.M., Jin, W., Shi, S.Q.: Interfacial debond of shape memory alloy composites. Smart. Mater. Struct. **14**, N29–N37 (2005)
18. Wang, X., Hu, G.K.: Stress transfer for a SMA fiber pulled out from an elastic matrix and related bridging effect. Compos. Part A **36**. 1142–1151 (2005)
19. Hu, Z.L., Xiong, K., Wang, X.: Study on interface failure of shape memory alloy (SMA) reinforced smart structure with damages. Acta. Mech. Sinica. **21**, 286–293 (2005)
20. Wang, X.M.: Shape memory alloy volume fraction of pre-stretched shape memory alloy wire-reinforced composites for structural damage repair. Smart Mater. Struct. **11**, 590–595 (2002)
21. Burton, D.S., Gao, X., Brinson, L.C.: Finite element simulation of a self-healing shape memory alloy composite. Mech. Mater. **38**, 525–537 (2006)
22. Tanaka, K.: A thermomechanical sketch of shape memory effect: one-dimensional tensile behavior. Res. Mech. **18**, 251–263 (1986)
23. 村上敬宜: 任意形状の板き　裂の応力拡大系数の計算法,日本機械学会论文集 (第1部),43卷370号(昭52-6),2022–2031 (1977)

Micromechanical Analysis of Mode I Crack Growth in Carbon Fibre Reinforced Polymers

Daniel Trias and Pere Maimí

Abstract Computational micromechanics models offer the possibility of analysis and quantification of the failure mechanisms that take place at the micro-scale level and are the responsible of the damage in the composite with high accuracy and with the need for very few hypotheses. Although these kind of analyses are common in current scientific literature, the analysis performed are generally limited to the stress/strain fields. This work makes use of a micromechanical model to analyze the crack tip and the cohesive zone of an interlaminar crack loaded in mode I for a carbon fiber reinforced polymer (CFRP). A periodic square fibre distribution is assumed and modelled in a FE environment and a degradation law is used to simulate damage in the matrix. This simulation allows both stress and strain quantification during crack opening and fracture mechanics analysis, such as the estimation of the critical value of the energy release rate and the quantification of the length of the cohesive zone, which is a parameter required for the application of cohesive elements.

1 Introduction

Delamination or interlaminar crack propagation is one of the most feared damage mechanisms in composite laminates since it reduces the stiffness and compressive strength of the material and it is difficult to detect.

D. Trias (✉) · P. Maimí
Analysis and Advanced Materials for the Structural Design (AMADE),
University of Girona, Girona, Spain
e-mail: dani.trias@udg.edu

The numerical simulation of delamination is generally conducted by using specific techniques such as the Virtual Crack Closure Technique (VCCT) [21, 22] or cohesive zone models [1, 7, 8, 26] but also mesomodels have been used sueccesfully to this end [23]. All these approaches deal with the homogenized composite and most of them define in advance the existence of a surface where interlaminar cracks are to grow. Moreover, the mesh dependence of the employed method must be analyzed prior to any structural simulation. While the mesh dependence of the VCCT comes from the self-similarity of the crack growth, any damage model used to simulate crack growth must ensure that the failure process zone is meshed with enough elements. Only a few works in the scientific literature tackled the simulation of delamination from the micromechanics' scope [9, 10] during the late 1980s. However, computational resources have changed enormously since then and a computational micromechanical model is able today to give much useful information.

Although the high stiffness and strength of laminated composites is due to the fibers, some potential dangerous damage forms as compressive fiber kinking, matrix cracking and delamination are promoted by the matrix properties. For this reason, a micromechanical model used to simulate delamination must include a reliable constitutive model for the matrix.

Epoxy resins are widely used as matrix phase in laminated composites reinforced with carbon, glass or other fibers. Polymer resins offer good specific properties to be used for aeronautical industries, the properties of epoxy offer a better adhesive, mechanical and hygroscopic properties compared with polyesters and vinyl-esters. The mechanical properties of epoxy are a key factor to understand the material response and the damage mode in the ply. This relation is not obvious due to the inhomogeneous stress field in the constituents under mechanical and thermal loads caused by the mismatch between the stiffness of fibers and matrix. The main result of this inhomogeneous stress field is the decrease of epoxy ductility when fibers are embedded in. Moreover, the inclusion of the fibers produces an increase of local strains respect to the homogeneous strain (the part of the material in serial respect to the loads) and a magnification of the stresses due the concentration of loads beside the fibers. Furthermore, the thermal variations in the curing process produces a residual stress states in the material. All this phenomena explain only partially the decrease of the matrix ductility in composite materials.

Although the wide variety of epoxy resins, it is possible to describe the main trend of the non-linear response of epoxies due to the type of applied load. The ductility of the epoxy resin decreases when the triaxial component of stresses acting on it increases. Under deviational stresses, when shear stresses are dominant, a large non-linear response with small hardening and permanent strains is observed. On the other hand, under hydrostatic stresses, where the pressure is dominant, the material fails brittle due to microcavitation. The experimental results by Asp et al. [3, 4] suggest that cavitation is obtained at constant dilatational energy, independently of the temperature. Even plastic flow is activated under distortional stresses the hydrostatic pressure plays and important role as the

Table 1 Toughness transfer from epoxy resin to composite from selected sources

Resin	Fibre	Resin type	Fibre volume (%)	G_{Ic} neat resin (N/mm)	G_{Ic} composite (N/mm)	Source
5208	T-300	Brittle epoxy	65	0.08	0.1	[18]
3501-6	AS-4	Brittle epoxy	67	0.095	0.175	[18]
HX205	T-300	Brittle epoxy	58	0.23	0.38	[18]
HX205	C-6000	Brittle epoxy	56	0.23	0.79	[18]

plastic flow is delayed when compressive pressures are present. Liang et al. experiments on pure epoxy show a linear increase of shear strength under compressive hydrostatic pressure: $\sigma_S = 31 - 0.12p$ [24]. These experiments suggest that the Drucker-Prager or the Mohr-Coulomb yield surface may be good candidates to predict the onset of plastic flow in pure epoxy. Interesting results were obtained by Fiedler et al. [13] which carried on a torsional load in a cylindric specimen, after a large amount of plastic response the failure plane is orientated in the principal tension stress direction as typical in brittle materials and not in shear as is typical in metals.

In spite of this general trend in the behavior of epoxies, different epoxy systems may have a different response with the same applied load, for example the experiments by Asp et al. [2–4] with DGEBA (diglycidyl ether of bisphenol) cured with DETA (diethylenetriamine) or APTA (polyoxy propyleneamine) show an inelastic regime before failure under an uniaxial tension load, specially if cured with APTA agent. On the other hand, the TGDDM (tetraglycidyl-4,4'-diaminodiphenyl methane epoxy) cured by DDS agent (4,4'-diaminodiphenyl sulphone) fails brittlely under uniaxial tension. When the specimen is loaded in biaxial tension DGEBA–APTA system appears to be the only specimen that yields prior to brittle failure.

Another open issue is the toughness transfer from matrix to the composite. As shown in Table 1, form similar epoxies and fibre volume, G_{Ic} may increase by a factor from 1.25 (T-300/5208) to 1.85 (AS-4/3501-6). Also, the effect of the fibre type for the same matrix should be analyzed since for the same matrix, the toughness transfer factor may change from 1.65 (T-300/HX205) to 3.44 (C-6000/HX205).

The aim of this work is to analyze several factors such as resin toughness and fibre Young Modulus which may affect the value of the toughness in a unidirectional composite under mode I crack growth. Up to this, a finite element model for the crack tip of a Double Cantilever Beam (DCB) specimen has been created. The model includes a with a simple damage model to simulate crack growth. Next section presents the used finite element model. Then, the details of the damage model are given. Finally the results of the simulation are given and some conclusions on the effect of constituents properties on G_{Ic} are drawn.

Fig. 1 Typical DCB specimen. δ denotes displacement control at tips. a0: pre-crack length. XFPZ: length of the failure process zone

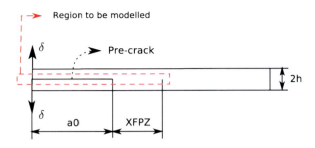

2 Finite Element Model

A micromechanics finite element model is constructed with the aim to simulate the cohesive damage zone located next to the crack tip in a DCB specimen [5, 20], schematized in Fig. 1.

The model assumes fibres are equally spaced forming an square array. The RVE is surrounded by an homogeneous material with the same elastic properties and thickness as the laminate in the specimen. However, to reduce the number of elements in this region and considering that this homogeneous material is only used to reproduce the same strain and stress fields present in the crack tip of a DCB specimen, the thickness has been reduced, changing the properties of the laminate to ensure it delivers the same flexure behaviour. In those outer planes perpendicular to the crack tip, periodicity conditions are employed, in the aim to simulate not only a fraction of the laminate, but a whole laminate. The fibre is considered to be a perfect elastic material with no failure or damage, since no fibre breakage is observed in tests. The matrix material is modeled with a damage law described in Sect. 2.1.

To correctly simulate the crack growth in the epoxy, the model must be able to include the whole Failure Process Zone. The size of the Failure Process Zone may be estimated using the following expression [27]:

$$\chi_{\text{FPZ}} = \alpha \frac{E'_I G_{Ic}}{\sigma_u^2} \quad (1)$$

where G_{Ic} is the mode I fracture toughness, σ_u is the strength and E'_i may be computed from the coefficients a_{ij} of the compliance matrix:

$$E'_I = \left(\frac{a_{11}a_{22}^{-\frac{1}{2}}}{2}\right)\left[\left(\frac{a_{22}}{a_{11}}\right)^{-\frac{1}{2}} + \frac{2a_{12} + a_{66}}{2a_{11}}\right]^{-\frac{1}{2}} \quad (2)$$

On the other hand, the coefficient α has been proposed to have the values of, among others: $\frac{2}{3\pi}$ [17], $\frac{1}{\pi}$ [19], $\frac{\pi}{8}$ [6, 11], $\frac{9\pi}{32}$ [12, 25] and 1.0 [16].

Consequently, the length of the cohesive zone might be somewhere between the proposed smaller and larger values for the coefficient $\alpha, \frac{2}{3\pi} = 0.21$ [17] and 1.0 [16].

In order to simulate the stress and strain fields in the crack tip of a DCB specimen without increasing the number of elements in the model, the arms of the

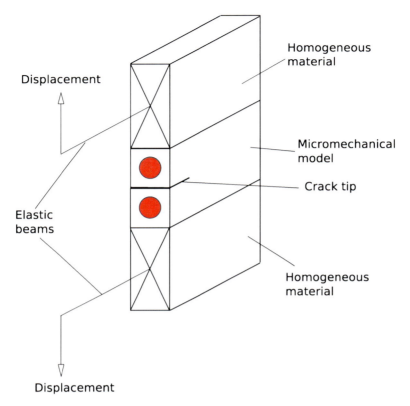

Fig. 2 Scheme of the finite element model. Displacement control is applied to the tips of two beams, which have the same elastic properties as the laminate, to reduce the number of elements in the model

DCB specimen are modelled using beams with the same elastic properties than the laminate. Displacements are applied at the tip of the beams, as sketched in Fig. 2.

2.1 Matrix Damage Model

The thermodynamics of irreversible processes is a general framework that can be used to formulate constitutive equations. It is a logical framework for incorporating observations and experimental results and a set of rules for avoiding incompatibilities. A constitutive damage model that has its foundation in irreversible thermodynamics is presented here. Since a finite element model with a large number of elements is employed, the constitutive model is simple. Based on the results by Asp et al. [3, 4], the failure envelop is based on the dilatational energy density. This way an explicit constitutive model, which guarantees fast computation, is developed.

2.1.1 Helmholtz Free Energy, Stress and Rate of Dissipation

The Helmholtz free energy is postulated to be of the form:

$$\Psi = \frac{1}{2} K \left((1-d) <\varepsilon_{ii}>^2 + <-\varepsilon_{ii}>^2 \right) + (1-d) G e_{ij} e_{ij} \quad (3)$$

where K and G are the bulk and shear elastic modulus, respectively. d is a scalar damage variables. The volumetric part of the damage variable is only activated under positive pressures to take into account the closure of cracks under compressive loads. The McCauley operator is defined as $<x> = (x+|x|)/2$ and e_{ij} is the deviational part of the strain tensor.

The rate of dissipation is defined as the difference between the mechanical power introduced $\sigma_{ij} \dot{\varepsilon}_{ij}$ and the variation of the Helmholtz free energy: $\Xi = \sigma_{ij} \dot{\varepsilon}_{ij} - \dot{\Psi} \geq 0$. Following the Coleman [28] arguments the constitutive response is defined as:

$$\sigma_{ij} = \frac{\partial \Psi}{\partial \varepsilon_{ij}} = K((1-d)<\varepsilon_{kk}> - <-\varepsilon_{kk}>)\delta_{ij} + 2(1-d) G e_{ij} \quad (4)$$

and the rate of dissipation: $\Xi = Y\dot{d} \geq 0$, where the thermodynamical force conjugated to the damage variable is defined as:

$$Y = -\frac{\partial \Psi}{\partial d} = \frac{1}{2} K <\varepsilon_{ii}>^2 + G e_{ij} e_{ij} \quad (5)$$

the positive dissipation is ensured if the damage variable is an increasing function.

The failure surface is considered as the maximum dilatational energy, according to the works by Asp [2–4]:

$$F_D = \sqrt{\frac{K}{2Y_K}} <\varepsilon_{kk}> - 1 - r_D \leq 0 \quad (6)$$

where Y_K is a material property (the critical dilatational energy density) which, if under an uniaxial test the material fails brittly, it can be related with the uniaxial strength as: $Y_K = \sigma_u^2/(18K)$. Experimental results by Asp et al. suggest that epoxy fails brittly under a constant hydrostatic density of energy. This value is almost constant at any temperature [3]. r_D is the internal variable of the model that evolves following the Khun–Tucker conditions, applying the consistence condition it can be integrated explicitly as:

$$r_D = \max_{s=0,t} \left\{ \sqrt{\frac{K}{2Y_K}} <\varepsilon_{kk}^s> - 1 \right\} \quad (7)$$

The damage variable is a monotonous increasing function of r_D that takes $d(r_D = 0) = 0$:

$$d = \min\left\{ 1, \frac{EG_C}{EG_C - 9KY_K \ell} \frac{r_D}{1+r_D} \right\} \quad (8)$$

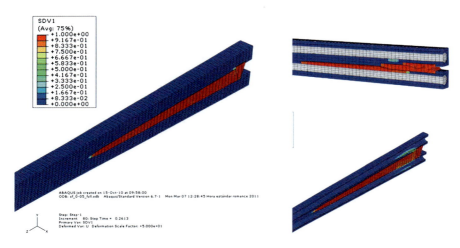

Fig. 3 Plot of the damage variable

This expression results in a lineal softening law with a density on energy dissipated as: G_C/ℓ under uniaxial stress. Where G_C is the fracture toughness and ℓ is the characteristic length of the finite element. To avoid snap-back in the constitutive response the element must be smaller than: $\ell \leq EG_C/(9KY_K)$.

The material is completely damaged when $\varepsilon_{ii}^F = \sqrt{\frac{2Y_K}{K} \frac{EG_C}{9KY_K\ell}}$, or $r_D^F = \frac{EG_C - 9KY_K\ell}{9KY_K\ell}$. The fracture toughness depends on the deviatonal load applied in the material.

$$G_C^{U_D/U_V} = \left(1 + \frac{U_D}{U_V}\right)\frac{EG_C}{9K} = \left(1 + \frac{U_D}{U_V}\right)\frac{1-2\nu}{3}G_C \tag{9}$$

The critical fracture energy under biaxial, triaxial stress and plane strain are $G_C^{BS} = (1-\nu)/2G_C$, $G_C^{TR} = (1-2\nu)/3G_C$, $G_C^{PS} = (1-\nu)/(1+\nu)G_C$, respectively.

The tangent operator is required to obtain a good convergence: $\dot{\sigma}_{ij} = C_{ijkl}^T \dot{\varepsilon}_{kl}$. When damage evolves the tangent operator is defined as:

$$C_{ijkl}^T = (1-d)C_{ijkl} - \frac{EG_C}{EG_C - 9KY_K\ell}\sqrt{\frac{2Y_K}{K}}(\varepsilon_{ss})^{-2} C_{ijmn}\varepsilon_{mn}I_{kl} \tag{10}$$

3 Results and Discussion

Model has been run for the different cases summarized in Table 2. As shown in Fig. 3, in all of the analyzed cases crack grows in the matrix/fibre interface, as observed experimentally [15].

Table 2 Combinations of crack propagation governing properties simulated

Case	Y_K(MPa)	G_{Ic} (N/mm)
1	0.1	0.01
2	0.1	0.05
3	0.1	0.1
4	0.1	0.5
5	0.1	1.0
6	0.2	0.01
7	0.2	0.05
8	0.2	0.1
9	0.2	0.5
10	0.2	1.0
11	0.5	0.01
12	0.5	0.05
13	0.5	0.1
14	0.5	0.5
15	0.5	1.0

3.1 Computation of Gc

Since force displacement pairs together with the crack length are available from the finite element model, G_{Ic} may be computed as if an experimental test was carried out, that is, by applying the procedure described in the ASTM procedure [5]. On the other hand, a theoretical expression for the force-displacement curve may be plotted using the expression [5]:

$$\delta = \frac{8a^2}{bh^3 E_{11}} \sqrt{\frac{G_{Ic} b^2 h^3 E_{11}}{12}} \qquad (11)$$

Force versus displacement curves are obtained for different values of G_I^c and Y_K. These curves are compared with the LEFM curve for the DCB test.

Acknowledgments The authors acknowledge the funding provided by the Spanish Ministry of Science and Innovation (MICINN) through research projects GRINCOMP (reference MAT2003-09768-C03-01) and EVISER2 (reference DPI2009-08048).

References

1. Allix, O., Corigliano, A.: Modeling and simulation of crack propagation in mixed-modes interlaminar fracture specimens. Int. J. Fract. **77**(2), 111–140 (1996)
2. Asp, L.E., Berglund, L.A., Gudmunson, P.: Effects of a composite-like stress state on the fracture of epoxies. Compos. Sci. Technol. **53**(1), 27–37 (1995)
3. Asp, L.E., Berglund, L.A., Talreja, R.: A criterion for crack initiation in glassy polymers subjected to a composite-like stress state. Compos. Sci. Technol. **56**(11), 1291–1301 (1996)
4. Asp, L.E., Berglund, L.A., Talreja, R.: Prediction of matrix-initiated transverse failure in polymer composites. Compos. Sci. Technol. **56**(9), 1089–1097 (1996)

5. American Standard Test Methods (ASTM).: Standard Test Method for Mode I Interlaminar Fracture Toughness of Unidirectional Fiber-Reinforced Polymer Matrix Composites. ASTM D5528-01 (2001)
6. Barenblatt, G.: The mathematical theory of equilibrium cracks in brittle fracture. Adv. Appl. Mech. **7**, 55–129 (1962)
7. Borg, R., Nilsson, L., Simonsson, K.: Modeling of delamination using a discretized cohesive zone and damage formulation. Compos. Sci. Technol. **62**(10-11), 1299–1314 (2002)
8. Chen, J., Crisfield, M., Kinloch, A.J., Busso, E.P., Matthews, F.L., Qiu, Y.: Predicting progressive delamination of composite material specimens via interface elements. Mech. Compos. Mater. Struct. **6**(4), 301–317 (1999)
9. Crews, J.H., Shivakumar, K.N., Raju, I.S.: Factors influencing elastic stresses in double cantilever beam specimens. NASA Technical Memorandum 89033 (1986)
10. Crews, J.H., Shivakumar, K.N., Raju, I.S.: A fiber-resin micromechanics analysis of the delamination front in a DCB specimen. NASA Technical Memorandum 100540 (1988)
11. Dugdale, D.S.: Yielding of steel sheets containing slits. J. Mech. Phys. Solids **8**, 100–104 (1960)
12. Falk, M.L., Needleman, A., Rice, J.R.: A critical evaluation of cohesive zone models of dynamic fracture. J. Phys. **IV**, 543–550 (2001)
13. Fiedler, B., Hojo, M., Ochiai, S., Schulte, K., Ando, M.: Failure behavior of an epoxy matrix under different kinds of static loading. Compos. Sci. Technol. **61**(11), 1615–1624 (2001)
14. Gdoutos, E.E.: Fracture Mechanis Criteria and Applications. Kluwer, Dordrecht (1990)
15. Greenhalgh, E.S.: Delamination growth in carbon-fibre composite structures. Compos. Struct. **23**(2), 165–175 (1993)
16. Hillerborg, A., Modéer, M., Petersson, P.E.: Analysis of crack formation and crack growth in concrete by means of fracture mechanics and finite elements. Cem. Concr. Res. **6**, 773–782 (1976)
17. Hui, C.Y., Jagota, A., Bennison, S.J., Londono, J.D.: Crack blunting and the strength of soft elastic solids. Proc. R. Soc. Lond. Ser. A. **459**, 1489–1516 (2003)
18. Hunston, Moulton et al.: Matrix resin effects in composite delamination. In: Johnston N.J. (ed.) Toughned Composites. ASTM STP vol. 937, pp. 74–94. ASTM International, West Conshohocken, Philadelphia, USA (1987)
19. Irwin, G.R.: Plastic zone near a crack and fracture toughness. In: Proceedings of the Seventh Sagamore Ordnance Materials Conference, vol. 4, pp. 63–78. Syracuse University, New York (1960)
20. International Standards Organization (ISO).: Standard test Method for Mode I Interlaminar Fracture Toughness, GIC, of Unidirectional fibre-reinforced Polymer Matrix Composites. ISO 15024 (2001)
21. Krüger, R.: The Virtual Crack Closure Technique: History, Approach and Applications. NASA/CR-2002-211628. NASA/CR-2002-211628 (2002)
22. Krüger, R., Cvitkovich, M.K., O'Brien, T.K., Minguet, P.J.: Testing and analysis of composite skin/stringer debonding under multi-axial loading. J. Compos. Mater. **34**(15), 1263–1300 (2000)
23. Ladevéze, P., Allix, O., Gornet, L., Lèveque, D.: A computational damage mechanics approach for laminates: identification and comparisons with experimental results. In: Voyiadkjis, G.Z., Ju, J.W.W., Chaboche, J.L. (eds.): Damage Mechanics in Engineering Materials, pp. 481–499. Elsevier Science B.V., North Holland (1998)
24. Liang, Y.-M., Liechti, K.M.: On the large deformation and localization behavior of an epoxy resin under multiaxial stress states.. Int. J. Solids Struct. **33**(10), 1479–1500 (1996)
25. Rice J.: The mechanics of earthquake rupture, physics of the earth's interior. In: Dziewonski, A.M., Boschi, E. (eds.) Proceedings of the International School of Physics Enrico Fermi, Course 78, 1979, pp. 555–649, Italian Physical Society and North-Holland (1980)

26. Turon, A., Camanho, P.P., Costa, J., Dávila, C.G.: A damage model for the simulation of delamination in advanced composites under variable-mode loading. Mech. Mater. **38**(11), 1072–1089 (2006)
27. Turon, A., Costa, J., Camanho, P., Maimí, P.: Analytical and Numerical Investigation of the Length of the Cohesive Zone in Delaminated Composite in Mechanical Response of Composites, pp. 77–97. Springer, Heidelberg (2008)
28. Maugin, G. A.: The Thermomechanics of Nonlinear Irreversible Behaviors. An Introduction. World Scientific Publishing, London (1999).

Modeling of Spalling Effect on Toughening in Fiber Reinforced Composites

C. Wang, N. F. Piaceski and K. M. Soares

Abstract According to statics and geometry of a random fiber crossing the cracked matrix surfaces, this work derived, at first, linear distribution of pressure provoked by the inclined fiber in the matrix and related the pressure to fiber axial force, bending moment and shear. To determine this pressure, self-consistent deformation model was proposed. With presuppose of constant interface fiber/matrix stress, the fiber and matrix displacements near the intersection of the fiber and the crack surfaces were calculated by means of integrating the Kelvin's fundamental solution of and the Mindlin's complementary solution, over a crack open of 0.001 mm. Then the stresses at points underneath of the inclined fiber were obtained using the same approach. The maxima normal stress in the fiber was calculated and compared with the fiber strength. Using failure criterion of five parameters for brittle materials, the spalling extent was determined. In the simulation, mechanical parameters of steel fiber and concrete were utilized and the percentages of active fibers were obtained with varying fiber strength and interface resistance. The results show that the percentage of active fibers increases with the fiber strength enhancement but decreases with the interface resistance increment. It draws a conclusion that the higher fiber strength and the lower interface resistance, more benefit to the debonding process. This work demonstrates that spalling effect is of great importance for fiber toughening in brittle matrix and the presented model allows for optimization of the parameters involved in toughening analysis.

C. Wang (✉)
Federal University of Pampa, Alegrete-RS 97546-550, Brazil
e-mail: wangchong@unipampa.edu.br

N. F. Piaceski · K. M. Soares
Northwest Regional University of Rio Grande do Sul, Ijuí-RS 98700-000, Brazil
e-mail: neivapiaceski@hotmail.com

K. M. Soares
e-mail: cameracolor@yahoo.com.br

Keywords Fiber reinforced composite · Toughening · Spalling effect · Bridging stress · Fiber/matrix · Interface · Kelvin's fundamental solution

1 Introduction

Brittleness of ceramics can be improved through adding ductile and strong fiber in the matrix. As too many parameters are involved in toughening, computational modeling is an economic and efficient approach to obtain the parameters optimization. For this objective, many computational models have been suggested [1–4]. However those models focused attention on the processes such as debonding interface between fiber and matrix, fiber put-out and snubbing effect. Few works of spalling analysis were reported [5]. To investigate the importance of matrix spalling, this work began from supposing the linear distribution of pressure provoked by a random inclined fiber in the matrix and then related the pressure to fiber axial force, bending moment and shear. The pressure was determine in terms of self-consistent fiber and matrix deformation and integration technique, which was based on the Kelvin's fundamental solution of and Mindlin's complementary solution. In the simulation section, the maxima normal stress in the fiber was calculated and compared with the fiber strength. Using failure criterion of five parameters for brittle materials, the spalling extent was determined. The percentages of active fibers (case for the fiber was not broken nor the spalling extent overpass the embedded length of the fiber) were computed with the variations of the fiber strength and the interface resistance.

2 Mathematical and Mechanical Modeling of Toughening by Fibers

2.1 Geometry of a Fiber Inclined to Crack Surfaces

Figure 1 illustrates the layout of a random fiber inclined to crack surfaces. The fiber is inclined to the crack surfaces at the angle θ with the surface normal. The embedded length of the short part of the fiber is noted as L_e while the full length as L_f and d_f the fiber diameter. We suppose that the initial debonded interface length is $2l_o$. The reason for this hypotheses is that the interface would be debonded in this part at very beginning when the two crack surfaces tend to move oppositely.

From geometry we have

$$L_e = \frac{L_f}{2} - \frac{z}{\cos\theta} - l_o \qquad (1)$$

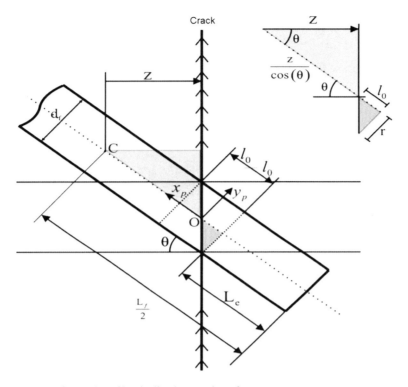

Fig. 1 Layout of a random fiber inclined to crack surfaces

where z is the distance from the fiber to the surface. By $l_0 = r \tan \theta$ and $L_e > 0$ and $z > 0$, we can get

$$0 < z < (L_f \cos \theta - d_f \sin \theta)/2 \quad \text{and} \quad \theta < \arctan(L_f/d_f) \tag{2}$$

2.2 Bridging Stress Provided by the Fibers Crossing the Crack Surfaces

There are a lot of fibers that cross the crack surfaces randomly. Each fiber contributes itself to link the two opening surfaces of the crack and intends to keep them from separating. The medium effect of the fibers is bridging stress. Just due to this bridging stress, the toughening is achieved. The bridging stress can be determined by the following equation [1]:

$$\sigma_c(w) = \frac{V_f}{A_f} \int_0^{\arctan(L_f/d_f)} \int_0^{(L_f\cos(\theta)-d_f\sin(\theta))/2} N(w,\theta,z)P(\theta)p(z)dzd\theta \qquad (3)$$

where V_f—volume fraction of fibers; A_f—section area of a fiber $N(w,\theta,z)$—the fiber axial force; w—the opening of the surfaces; $P(\theta)$ and $p(z)$—the distribution functions of fiber inclined angles and fiber centers respectively, given by $P(\theta) = \sin(\theta)$ for $0 \leq \theta \leq \arctan(L_f/d_f)$ and $p(z) = 2/L_f$ for $0 \leq z \leq (L_f\cos(\theta)-d_f\sin(\theta))/2$.

By Eq. 3 we see that to obtain the bridging stress the key is to find the axial force of the inclined fiber $N(w,\theta,z)$ in relation to the crack open w and the embedded length L_e or the distance of the fiber center to the surface z. This is the major difficulty in toughening simulation, because fiber toughening involves two distinct processes: interface debonding and fiber pull-out. For the former, there occurs spalling effect and, for the later snubbing effect will come into action.

2.3 Increment of Fracture Energy G_c Contributed with Fibers

After obtaining the bridging stress the increment of fracture energy G_c (J/m^2) i.e. toughening can be calculated as the area under the curve $\sigma_c - w$, using the following integral:

$$G_c = \int_0^{w^*} \sigma_c(w)\,dw \qquad (4)$$

where w^* is the ultimate crack open for which all fiber axial forces become zero, that is, all the fibers are pulled apart or out the matrix.

2.4 Spalling in Matrix

When a closed crack is opening, the fiber imposes pressure on the matrix. Before debonding occurs, this pressure goes up to the maximum. It is possible that the stresses underneath the inclined fiber near the point where the fiber exists from the matrix provoked by the pressure is higher than the matrix strength, then matrix spalling comes out (Fig. 2). If the extent of spalling overpasses the embedded length of a fiber, this fiber will lose its contribution to toughening; otherwise the relieved stresses rid the fiber of possibility to be broken, in other word, the fiber is saved and can make contribution to toughening.

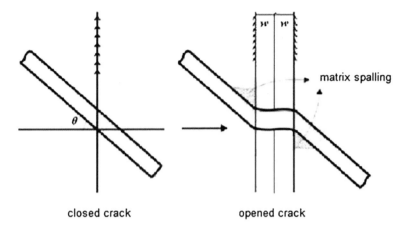

Fig. 2 Spalling effect

2.5 Analysis of Displacement and Deformation of an Inclined Fiber with the Crack Opening

From Eq. 3 we see that to obtain the bridging stress, the key is to find the axial force of the inclined fiber $N(w, \theta, z)$ in relation to the crack open w and the embedded length L_e or the distance of the fiber center to the surface z. This is the major difficulty in toughening simulation, because fiber toughening involves two distinct processes: interface debonding and fiber pull-out.

When the crack surfaces open, the inclined fiber will suffer deformation (Fig. 3). The bridging force F_{br} of the fiber is the resultant of the axial force, N and the shear, P and perpendicular to the crack surfaces. The axial force, N and the shear, P lead to the displacements u_1 and u_2 of the section center of the embedded part of the fiber at the point where the fiber exits the matrix and the deformation of the unembedded part (the debonded part $2l_o$, Fig. 4). It is easy to know:

$$F_{br} = N \cos \theta + P \sin \theta = N / \cos \theta \quad (5)$$

The stretched length Δl of the embedded fiber has a component in the direction normal to the crack surfaces i.e. in the direction u_1 (Fig. 4). Therefore we can write the half open of the crack as

$$w = \Delta l \cos \theta + u_1 \quad (6)$$

where the stretch $\Delta l = Nl_0/A_f E_f$.

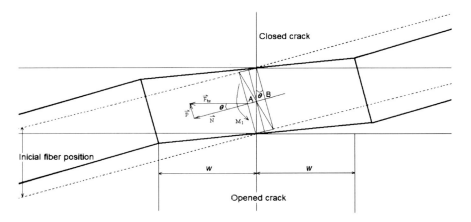

Fig. 3 Fiber deformation and interior forces after the crack opened to 2w. N—axial force, P—shear and M—moment are on the fiber section at the point where the fiber exits the matrix

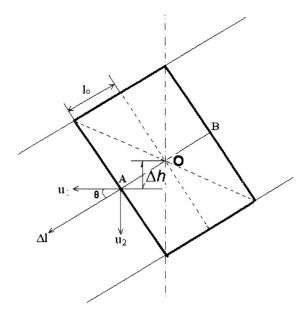

Fig. 4 The displacements, u_1 and u_2 and the deformation, Δl of the debonded part of the fiber attributed to the axial force, N and the shear, P; Δh is the vertical distance between the points A and O

2.6 Pressure Imposed by Fiber on Matrix

The action of the fiber interior forces N, P and M will produce a pressure onto the matrix (Fig. 5). This pressure is unknown completely. For the possibility to resolve the problem we suppose a linear distribution for the pressure as

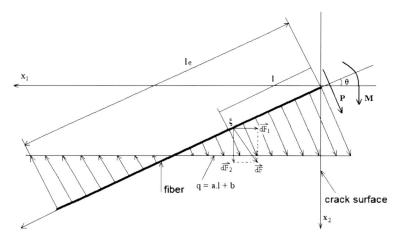

Fig. 5 Fiber pressure onto the matrix

$$q = al + b \qquad (7)$$

where a and b are unknown constants.

By static equilibrium

$$P = \int_0^{le} q\,dl = \int_0^{le} (al+b)\,dl \quad \text{and} \quad -M = \int_0^{le} lq\,dl = \int_0^{le} (al+b)l\,dl \qquad (8)$$

we can express a and b as functions of P and M

$$a = -6\left(\frac{2M}{le} + P\right)/l_e^2, \quad b = 6\left(\frac{2Pl_e}{3} + M\right)/l_e^2 \qquad (9)$$

The interface shear stress is determined as $q_a = N/d_f \pi l_e$.

As the debonded part of the fiber ($2l_0$ in Fig. 4) is anti-symmetrical about the origin O, the moment on the diagonal section coincident with the closed crack should be null. Hence we get the relationship

$$F_{br} \cdot \Delta h = M \qquad (10)$$

where M is the moment on the transversal section, A of the left part of the fiber. From geometry and the stretch, $\Delta l = N l_0 / A_f E_f$, we get

$$\Delta h = \Delta l.\sin\theta + u_2^e + r.\sin\theta.tg\theta = \left(\frac{N.l_o}{E_f A_f}\right)\sin\theta + u_2^e + r.tg\theta \sin\theta \qquad (11)$$

By Eqs. 10 and 11 we obtain

$$M = F_{br}\left(\frac{N.l_o.\sin\theta}{E_fA_f} + u_2^e + r.tg\theta\sin\theta\right)$$
$$= F_{br}\left[\left(\frac{N}{E_fA_f} + 1\right)r.tg\theta\sin\theta + u_2^e\right] \quad (12)$$

For full crack open, one can write

$$2w = u_1^e + u_1^d + \left(\frac{2N.l_o}{E_fA_f}\right).\cos\theta = u_1^e + u_1^d + \left(\frac{N.d_f}{E_fA_f}\right).\sin\theta \quad (13)$$

In such a way we relate the crack open with the axial force, N. At first appearance the crack open is only linked to the axial force, N but one should not forget that u_1^e and u_1^d are still functions of N, P, and M. In fact, Eq. 13 is self-consistent. For a given w, the correct N should satisfy Eq. 13. Since the force, N is tensile, then u_1^e and u_1^d should be negative. The displacements, u_1^e and u_1^d play a role in diminishing the crack open. From other side the component of the stretched part, $Nd_f/E_fA_f \sin\theta$ tries to open the crack more. There is competition between the two sides. It would occur that the crack open is null in spite of increasing N. Usually the sum of the displacements, u_1^e and u_1^d is very small compared to the stretch component $(Nd_f/E_fA_f)\sin\theta$, so we can ignore the sum and simplify the calculation as

$$2w \approx (Nd_f/E_fA_f)\sin\theta \quad (14)$$

Then

$$N \approx 2wE_fA_f/d_f \sin\theta \quad (15)$$

After N is determined, the other parameters can be resulted without difficulty. The parameters a and b become

$$a = -6F_{br}[\sin\theta + 2(F_{br}.\cos\theta.l_o.\sin\theta/E_fA_f + u_2^e + r.tg\theta.\sin\theta)/le]/le^2 \quad (16)$$

$$b = 6F_{br}[2\sin\theta.le/3 + Fbr\cos\theta.l_o.\sin\theta/E_fA_f + u_2^e + r.tg\theta.\sin\theta]/le^2 \quad (17)$$

2.7 Decomposition of Circumferential Pressure and Friction on the Fiber Surface

The pressure, q varies linearly with the length of the embedded fiber and has the unit (N/m). As the fiber section is circular, the linear pressure, q must be resultant of circumferential pressure, p having the unit (N/m^2). The pressure, p must not be

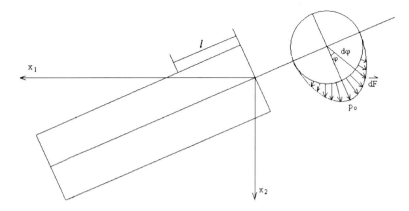

Fig. 6 The pressure, p distributed circumferentially

uniformly distributed around the fiber surface. The circumferential pressure, p can be assumed reasonably as

$$p = p_o \cos \phi \qquad (18)$$

where p_0 is the maxima pressure that just occurs at the bottom surface (Fig. 6). It can be seen that the pressure will be zero when the angle ϕ hits $-90°$ and $90°$.

It is easy to obtain

$$p_o = \frac{4(al+b)}{d_f \pi} \qquad (19)$$

Then

$$p = \frac{4(al+b)\cos\phi}{d_f \pi} \qquad (20)$$

The elementary force, dF caused by the pressure on the elementary area of the fiber surface can be decomposed as dF_3 and dF', which is in turn decomposed into dF_1 and dF_2 according to the coordinates x_1 and x_2 (Fig. 7). Then we get the elementary forces:

$$dF_1 = -dF' \sin\theta = -4(al+b)\cos^2\varphi \sin\theta \cdot dl \cdot r_f \cdot d\varphi \qquad (21)$$

$$dF_2 = dF' \cos\theta = 4(al+b)\cos^2\varphi \cos\theta \cdot dl \cdot r_f \cdot d\varphi \qquad (22)$$

$$dF_3 = -dF \sin\varphi = -2(al+b)\sin 2\varphi \cdot dl \cdot r_f \cdot d\varphi \qquad (23)$$

The elementary forces caused by the friction can be written as

$$dF_a = q_a \cdot dl \cdot ds = q_a \cdot r_f \cdot dl d\varphi \qquad (24)$$

Fig. 7 Decompositions of the elementary force and friction on the fiber surface

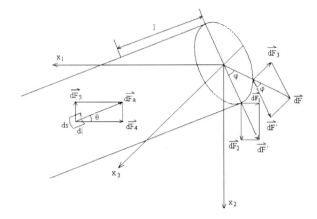

$$dF_4 = -dF_a \cos \theta = -q_a \cdot r_f \cdot \cos \theta \cdot dl d\varphi \qquad (25)$$

$$dF_5 = -dF_a \sin \theta = -q_a \cdot r_f \cdot \sin \theta \cdot dl d\varphi \qquad (26)$$

2.8 Kelvin's Fundamental Solution and Mindlin's Complementary Solution

The Kelvin's fundamental solution [6, 7] plays an important role in solid mechanics. Kelvin derived the analytic expressions of the displacements and stresses provoked by unit point loads in an *infinite* elastic medium. To keep the validation of the Kelvin's solution in a semi-infinite solid, Mindlin gave the additional solution, called the Mindlin's complementary solution [8]. So the displacement and stress solutions attributed to the action of unit point loads can be written respectively as

$$u_{ij} = u_{ij}^{(K)}(\xi, x) + u_{ij}^{(C)}(\xi, x) \qquad (27)$$

$$\sigma_{kji} = \sigma_{kji}^{(K)}(\xi, x) + \sigma_{kji}^{(C)}(\xi, x) \qquad (28)$$

$$u_{ij}^{(K)}(\xi, x) = \left\{ \frac{1}{16\pi(1-v)G}(3-4v)\delta_{ij} + r_{,i}.r_{,j} \right\}, \quad i,j,k = 1,2,3 \qquad (29)$$

$$\sigma_{jki}^{(K)}(\xi, x) = -\frac{1}{4\alpha\pi(1-v)r^\alpha}\left\{(1-2v)(r_{,k}\delta_{ij} + r_{,j}\delta_{ki} - r_{,i}\delta_{jk}) - \beta r_{,i}.r_{,j}.r_{,k}\right\} \qquad (30)$$

where $\alpha = 2, 1$ and $\beta = 3, 2$ for three- and two-dimensional plane strain respectively; v—Poisson ratio; G—shear elasticity modulus; r—the distance from the

Fig. 8 Unit point forces applied in a half space (after Brebbia et al. [7])

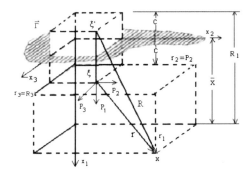

source point, ξ to the field point, $\mathbf{x}$; the Kelvin's solution, $u_{ij}^{(K)}(\xi, x)$ represents the displacement in the j-direction at a field point, $\mathbf{x}$ corresponding to a unit force acting in the i-direction applied at a source point, ξ; $\sigma_{jki}^{(K)}(\xi, x)$ are the stress tensor. Figure 8 depicts the geometric configuration for the case of a half space.

2.9 Determination of Stresses in the Matrix

To calculate stresses in the matrix, we adapt the coordinate system (x_1, x_2, x_3) with the origin O (Fig. 9). Then for a source point, $\xi = (r_f, \phi, x_1')$, the coordinates are

$$x_1(\xi) = r_f \sin\theta + x_1' \cos\theta - x_2' \sin\theta \tag{31}$$

$$x_2(\xi) = -r_f \cos\theta + x_1' \sin\theta + x_2' \cos\theta \tag{32}$$

$$x_3(\xi) = -r_f \sin\varphi \tag{33}$$

For any field point, x in the matrix where the stresses will be computed, we have

$$r_i = x_i(\mathbf{x}) - x_i(\xi) \tag{34}$$

$$r_1 = x_1 + r_f \sin\theta \cos\phi - l\cos\theta - r_f \sin\theta \tag{35}$$

$$r_2 = x_2 - r_f \cos\theta \cos\phi - l\sin\theta + r_f \cos\theta \tag{36}$$

$$r_3 = x_3 + r_f \sin\phi \tag{37}$$

$$r = \sqrt{r_1^2 + r_2^2 + r_3^2} \tag{38}$$

Fig. 9 Coordinate systems used

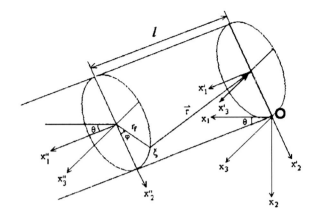

$$r_{,1} = \frac{\partial r}{\partial x_1} = \frac{r_1}{r} \tag{39}$$

$$r_{,2} = \frac{\partial r}{\partial x_2} = \frac{r_2}{r} \tag{40}$$

$$r_{,3} = \frac{\partial r}{\partial x_3} = \frac{r_3}{r} \tag{41}$$

For the image ξ' of the source point x the coordinates are

$$R_i = x_i(\mathbf{x}) - x_i(\xi') \tag{42}$$

$$R_1 = x_1 - r_f \sin\theta \cos\phi + l\cos\theta + r_f \sin\theta \tag{43}$$

Then the parameters involved in the Kelvin's fundamental solution and the Mindlin's complementary solution are given as

$$c = x_1(\xi) = -r_f \sin\theta \cos\phi + l\cos\theta + r_f \sin\theta \geq 0 \tag{44}$$

$$\bar{x} = x_1(\mathbf{x}) \geq 0 \tag{45}$$

In a half space the stresses in the matrix can be obtained using the integration over all of surface loads:

$$\sigma_{jk} = \iint_{F_s} d\sigma_{jk} = \iint_{F_s} (\sigma_{jki}^K + \sigma_{jki}^C) dF_i, \quad i = 1, 2, 3 \tag{46}$$

where σ_{jki}^K and σ_{jki}^C are the Kelvin's fundamental solution and the Mindlin's complementary solution.

Fig. 10 The polar coordinates net of the region supporting the fiber. At the nodes matrix damage were checked

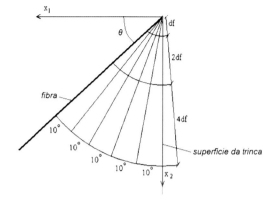

3 Simulation and Result Analysis

The parameters utilized in this work come from the experiment of cement reinforced by steel fibers [5]. Fantili and Vallini also used them for their research work [9]. These parameters are: $L_f = 10$ mm (fiber length), $d_f = 0,5$ mm (fiber diameter), $E_f = 200$ GPa (fiber elasticity modulus), $v_f = 0.3$ (fiber coefficient of Poisson), $E_m = 30$ GPa (matrix elasticity modulus), $v_m = 0.2$ (matrix coefficient of Poisson), $f_t = 3.7$ MPa (matrix tensile strength), $f_c = 36.5$ MPa (matrix compressive strength) and σ_f (fiber tensile strength) and τ_i (interface shear strength) vary.

The pressure attributed by an inclined fiber and imposed onto the matrix would result matrix spalling. To identify if the matrix is damaged, we have used the criterion of five parameters for brittle material. The Ref. [10] details this criterion.

The verification of spalling was taken only at some points in the compressive region that is supporting the fiber. We set them as nodes of a polar coordinate net (Fig. 10). The arc radii of the net expand forth with the values: df, 3df, 7df, 15df till to 1.5 times the embedded length l_e. The support angle (90°–θ) was divided into small ones of 10° approximately. To trace the evolution of spalling (damage), we applied the axial load N at several small steps till the fiber was broken i.e. the fiber tensile strength was reached, then, recorded the regions damaged respectively.

Figure 11 illustrates the spalling evolutions for different inclined angles but with much same fiber embedded length. The value in the legend indicates how much was the bridging force meanwhile the fiber was broken. The damage always is initiated at the point where the fiber exits the matrix. For the fiber of small inclined angle (Fig. 11a), the spalling region is small at beginning but expands rapidly. The fiber does not break until high bridging force is reached (about 40% of fiber strength in force $F_f = \sigma_f A_f$). However for the fiber of large inclined angle (Fig. 11b), the initial damage extent already is quite large. Only by a very low bridging force (only about 5% of the fiber strength in force F_f), the fiber has been

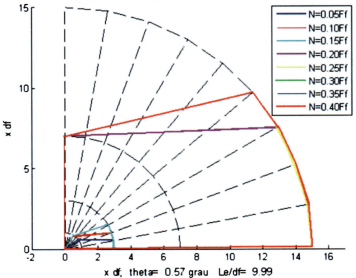
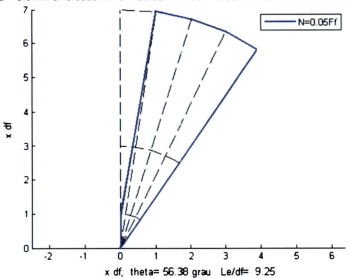

Fig. 11 Evolution of spalling. **a** θ = 0.57°; **b** θ = 56.38°

Table 1 Percent of fibers fortune under the conditions: $L_f = 10$ mm, $d_f = 0.5$ mm, $E_f = 200$ GPa, $v_f = 0.3$, $E_m = 30$ GPa, $v_m = 0.2$, $f_t = 3.7$ MPa, $f_c = 36.5$ MPa and $V_f = 5\%$ (volume fraction of fibers in the matrix)

σ_f/τ_i	Broken (%)	Debonded (%)	Damaging matrix (%)	Perfect (%)	Active (%)	$G_c(10^{-3} J/m^2)$
469/1	14	58	20	8	86	2.7
469/3	24	45	28	11	76	6.0
469/5	32	36	21	11	68	6.0
635/1	11	61	17	11	89	3.0
635/3	22	46	19	13	78	6.6
635/5	28	40	19	13	72	6.6
954/1	9	64	17	10	91	2.9
954/3	17	52	17	14	83	6.5
954/5	23	46	17	14	77	7.2

G_c toughening listed only for comparison

broken. From Eq. 12 it would be known that high inclination leads to large bending moment. As a consequence, the bending stress is large although the bridging force is still low.

In the last section, the analysis focuses attention on the evolution of damaged region but not on the fibers. In this section we will discuss the influence of the fiber strength, σ_f and the interface strength, τ_i on the percentage of alive fibers.

Spalling usually occurs before the initiation of interface debonding because the fiber can provide the maximum axial force. For this reason we set the crack open, w^* up to 0.001 mm for all simulations and divide it into 20 small increments. We also split the inclined angle, θ into small parts denoted by the number, nt and the embedded length by nz. In this way the number of fibers is $nt \times nz$. We set nt = 20 and nz = 10 in this work. That is said 200 fibers are involved. To identify if the matrix is damaged, we employ the criterion of five parameters for brittle material [10]. The fortune of a fiber went into four categories: broken—the stress in the fiber hit the fiber strength,σ_f, debonded—the maximum shear stress on the surface of the embedded fiber was equal to or larger than the interface strength, τ_i, damaging matrix—the extent of damage of the matrix overpassed the embedded length of the fiber and perfect—not included in the former three categories. The active fibers were those that have not been broken. The following table lists the computational results.

Table 1 reveals the importance of spalling because there are about 20% fibers that did damage to the matrix. If these fibers did not produce the damage, they would bear more load continuously and make more contribution to toughening. The majority of reported modeling works in computational toughening was on such a hypothesis. They obtained larger toughening than the one from experiment [11]. The importance of spalling also was confirmed by Pavan [12].

From Fig. 12 it is obvious that for given fiber strength, σ_f, the high interface strength, τ_i resulted in more fibers broken; and for given interface strength, high

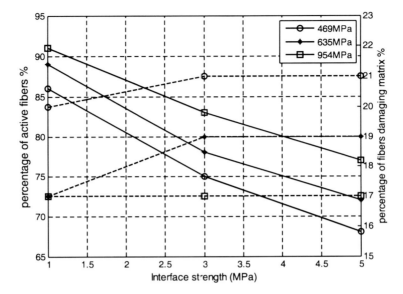

Fig. 12 Influence of fiber strength and interface strength on the percentage of active fibers and the percentage of fibers that lead the matrix damage—for active fibers—for fibers damaging matrix

fiber strength rid more fibers of being broken. Therefore, we draw a conclusion that high fiber strength and low interface strength benefit toughening.

4 Conclusion

We come to the conclusion that (1) spalling effect is far important for fiber toughening. Ignoring this effect would lead to an exaggerative toughening; (2) high fiber strength and low interface strength benefit toughening.

Acknowledgments The authors Wang and Soares acknowledge the National Council for Scientific and Technological Development (CNPq, an agency linked to the Ministry of Science and Technology, Federal Republic of Brazil) for its finance aid (Grant Numbers: 501660/2009-7 and 562244/2008-5) in this work.

References

1. Li, V.C., Wang, Y., Backer, S.: J. Mech. Phys. Solids **39**, 607 (1991)
2. Li, V.C.: Cement & Concrete Compos. **14**,131 (1992)
3. Lin, Z., Li, V.C.: J. Mech. Phys. Solids **45**, 763 (1997)

4. Wang, Y.: Int. J. Cement Compos. Lightweight Concrete **10**, 143 (1988)
5. Leung, C.K.Y., Shapiro, N.: J. Mater. Civil Eng. **11**,116 (1999)
6. Love, A.E.H.: Treatise on the Mathematical Theory of Elasticity. Dover, New York (1944)
7. Brebbia, C.A., Telles, J.C.F., Wrobel, L.C.: Boundary Element Techniques: Theory and Applications in Engineering. Computational Mechanics, New York (1984)
8. Mindlin, R.D.: Physics **7**,195 (1936)
9. Fantilli, P.A., Vallini, P.: J. Adv. Concrete Technol. **5**, 247 (2007)
10. Ansys, User's Manual—Theory, Vol. IV (SAS IP, 1994)
11. Maalej, M., Li, V.C., Sashida, T.: Effect of fiber rupture on tensile properties of short fiber composites. J. Eng. Mech **121**, 903 (1995)
12. Pavan, Alcione R.: Mathematical Modeling of ceramic thoughening by adding micro fibers. Master Paper, Northwest Regional University of Rio Grande do Sul, Ijuí, 2006. (in Portugues)

Evolution Characteristics of Delamination Damage in CFRP Composites Under Transverse Loading

S. S. R. Koloor, A. Abdul-Latif, X. J. Gong and M. N. Tamin

Abstract The initiation and subsequent progression of delamination in CFRP composite laminates is examined using finite element method. A 12-ply CFRP composite, with a total thickness of 2.4 mm and anti-symmetric ply sequence is simulated under three-point bend test setup. Each unidirectional composite lamina is treated as an equivalent elastic and orthotropic panel. Interface behavior is defined using cohesive damage model. Complementary three-point bend test on the specimen is performed at crosshead speed of 2 mm/min. The measured load–deflection response at mid-span location compares well with predicted values. Interface delamination accounts for up to 46.7% reduction in flexural stiffness from the undamaged state. Delamination initiated at the center mid-span region for interfaces in the compressive laminates while edge delamination started in interfaces with tensile flexural stress in the laminates. Anti-symmetric distribution of the delaminated region is derived from the corresponding anti-symmetric ply sequence in the CFRP composite. The dissipation energy for edge delamination is greater than that for internal center delamination.

S. S. R. Koloor · A. Abdul-Latif · M. N. Tamin (✉)
Faculty of Mechanical Engineering, Center for Composites,
Universiti Teknologi Malaysia, 81310 UTM Skudai,
Johor, Malaysia
e-mail: taminmn@fkm.utm.my

X. J. Gong
Département de Recherche en Ingénierie des
Véhicules pour l'Environnement,
Université de Bourgogne, 58000 Nevers, France

1 Introduction

Carbon fiber-reinforced polymer (CFRP) matrix composites are commonly employed in advanced structural applications including the skin of aircraft wings, stringers, rotor blades and molded beams. The increasing use of CFRP composites is derived from its high strength-to-weight ratio, high specific modulus and improved corrosion resistance compared to their metallic counterparts. The flexible design of laminated composites through sequencing of the pre-impregnated laminates enables tailoring of the composite structures to specific design and functions. Typical loading in these structures often involves lateral bending of the composite laminates, both transient and fatigue. Several damage and fracture scenarios of the composite panels under complex applied loading conditions should then be considered for optimum design of the structure. Thus, reliability of a CFRP composite structure under such load conditions depends on the continual process of damage initiation and subsequent damage propagation to catastrophic fracture.

Several modes of failure observed in CFRP composites include matrix cracking, fiber/matrix interface debonding, fiber fracture, fiber pull-out and interface delamination. The latter is of particular interest in view of the relatively weak ply-to-ply interface strength and the large interface shear stresses developed during loading. Interface delamination failure leads to significant loss of the load-carrying capacity and occurs in the absence of any visible damage of the CFRP composite. In practice, internal defects including microcracks, voids and interface delamination are detected employing non-destructive evaluation techniques such as ultrasonic testing and thermography. The progression of each of these initial processing damages under operating loads to failure can then be monitored and quantified. The critical delamination failure process could be accounted for in the design phase with the aid of numerical modeling and validation testing of CFRP composite specimens.

The mechanics of bi-material interfaces particularly in laminated composite materials and structures has been well studied (e.g. [6, 10]). An interface decohesion model was formulated using damage mechanics concept and used to examine the composite failure process by simulating delamination initiation and subsequent progression in fiber-reinforced composite [4]. In this cohesive zone model, the state of interface damage is described in terms of stress-to-strength ratios of normal and shear traction components in a quadratic failure criterion along with a mixed-mode displacement formulation for crack initiation event. Subsequent crack propagation is predicted based on fracture energy considerations [2, 7]. Limited work is available on the extension of the cohesive zone model to account for cyclic-induced damage of the interfaces in CFRP composites during fatigue loading.

In this study, the mechanics of interface delamination in CFRP composite is examined. Both cases with perfect non-damaging ply interfaces and damaging interfaces by delamination of CFRP composites under flexural loading are

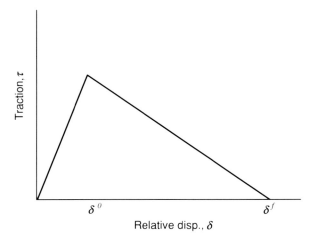

Fig. 1 Traction-displacement curve for idealized bilinear softening law

simulated. Interface delamination behavior is described using cohesive zone model. Validation of the model is judged through comparisons of predicted results and measured values of load–displacement response of the composite specimen.

2 Cohesive Zone Model

Key equations describing the cohesive zone model for interface delamination are reviewed here. Details of the cohesive element formulation are found elsewhere [4, 5]. A typical traction-relative displacement curve of a bi-material interface is illustrated in Fig. 1. As a first approximation, a bilinear traction-relative displacement softening law is employed to relate the traction, τ to the relative displacement, δ at the interface. In this model, the cohesive zone can still transfer the load after the onset of damage, δ^o. In pure debonding mode I (tensile opening), mode II (shear) or mode III (out-of-plane tearing) the onset of damage occurs when the normal or shear tractions attained their respective interface tensile or shear strength, respectively. The interface stiffness is then gradually reduced to zero when complete fracture occurs at δ^f. For small cohesive zone, the critical energy release rate, G_C for the respective fracture mode can be estimated by the area under the τ-δ curve which also represents the work of fracture as:

$$G_C = \int_0^{\delta^f} \tau(\delta)\, d\delta \tag{1}$$

Materials damage in the bi-material interface plane is quantified using a quadratic function of the orthogonal traction components. The debonding or delamination criterion for a material point under a mixed-mode loading is then prescribed as:

$$\sqrt{\left(\frac{\langle\sigma_{22}\rangle}{T}\right)^2+\left(\frac{\tau_{21}}{S}\right)^2+\left(\frac{\tau_{23}}{S}\right)^2} = 1 \qquad (2)$$

where $\langle\sigma_{22}\rangle$ is the normal traction to the interface plane defined as σ_{22} for non-zero value and zero otherwise, τ_{21} and τ_{23} are the transverse tractions. Orientations of the orthogonal axes (x, y, z) are as illustrated in Fig. 3 and corresponds to the subscripts (11, 22, 33), respectively. The terms T and S are the nominal tensile and shear strength of the interface, respectively. The mixed-mode damage initiation displacement, δ_m^o defines the relative displacement corresponding to the onset of crack initiation, taking into account the three Cartesian displacement components, δ_1^o and δ_3^o in the interface shear plane while δ_2^o normal to the plane, as follows:

$$\delta_m^o = \delta_1^o \delta_2^o \sqrt{\frac{1+\beta^2}{\left(\delta_1^o\right)^2+\left(\beta\delta_2^o\right)^2}} \qquad (3)$$

The term $\beta = \dfrac{\delta_{\text{shear}}}{\delta_2}$, $\delta_2 > 0$ represents mode mixity for non-zero relative opening displacement. In this equation, the same delamination mechanisms in mode II and mode III are assumed. Thus, δ_{shear} represents the total tangential displacement for the two orthogonal relative displacement components δ_1 and δ_3.

Once interface delamination has initiated, subsequent interface crack propagation under mixed-mode loading conditions is predicted in terms of total energy release rates, $G_T = G_I + G_{II}$ and single-mode fracture toughness, G_{IC} and G_{IIC} as [1]:

$$G_T = G_{IC} + (G_{IIC} - G_{IC})\left(\frac{G_{II}}{G_T}\right)^\eta \qquad (4)$$

The parameter, η describes the interaction of fracture modes. The numerical value of η can be obtained by best-fitting of experimental data for combinations of crack loading modes. Typical value for laminated composite samples is available [8].

3 Materials and Experimental Procedures

A CFRP composite laminate was laid up using high modulus carbon fibers and epoxy resin (M40J fibers and NCHM 6376 resin, Structil France). The composite was fabricated with anti-symmetric ply sequence of [45/-45/45/0/-45/0/0/45/0/-45/45/-45] with the 0°-direction along the length of the specimen. The 12-ply CFRP composite laminate with a total thickness of 2.4 mm is machined into specimens measuring 140 mm × 20 mm. The longitudinal cross section of the composite laminate is shown in Fig. 2 with each lamina having a thickness of 0.2 mm. Lamina with 0° fiber orientation are easily distinguishable as shown in the figure. Transient three-point-bend test is performed on the specimen with 112 mm-span

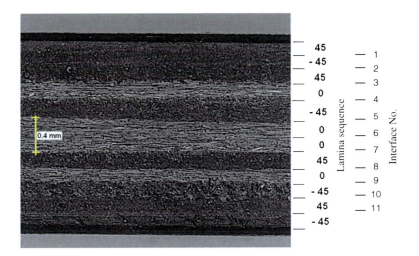

Fig. 2 Longitudinal cross section of CFRP composite

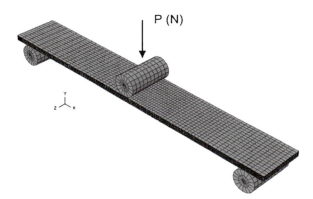

Fig. 3 Model of CFRP composite specimen under three-point bend test setup

and at machine crosshead speed of 2 mm/min. The test is conducted to a pre-scribed mid-span deflection of 18 mm. Load–deflection response of the specimen at the mid-span location are recorded throughout the test.

4 Finite Element Modeling

A typical finite element model of the CFRP composite specimen under the three-point-bend test set-up is illustrated in Fig. 3. Each unidirectional composite lamina is treated as equivalent elastic, orthotropic panel. Two different cases, one

Table 1 Elastic properties and materials data for unidirectional lamina and interface damage model of CFRP composite

Lamina constants		Fiber/matrix interface properties	
E_1	105.5 GPa	E	1.0×10^6 MPa
E_2	7.2 GPa	G_1	1.0×10^6 MPa
E_3	7.2 GPa	G_2	1.0×10^6 MPa
G_{12}	3.4 GPa		
G_{13}	3.4 GPa	Data for cohesive damage model	
G_{23}	2.52 GPa	$T = 26.12$ MPa	$G_{IC} = 0.26$ N/mm
v_{21}	0.34	$S = 14.57$ MPa	$G_{IIC} = 0.52$ N/mm
v_{31}	0.34	$\eta = 54.1$	$G_{IIIC} = 0.52$ N/mm
v_{32}	0.378	–	

representing perfectly bonded laminate interfaces and the other, degraded interfaces are modeled. The model of the composite with perfectly bonded interface is used to examine characteristics of shear stresses in each ply interface under flexural loading. In the composite with degraded interfaces, the response of the interface is described using a damaged, linear elastic constitutive model. The process-induced residual stress due to coefficient of thermal expansion mismatch among the different lamina lay-ups during curing of the composite is not considered. Each of the 0.2-mm thick ply is modeled with a single layer of finite elements. The lamina is modeled using 8-node solid elements. In the composite model with degraded interfaces, the ply-to-ply interfaces are discretized using cohesive interface elements with matching mesh to the solid elements on both surfaces. Non-uniform mesh density is employed along the composite specimen length.

The prescribed boundary conditions of the model represent the test setup. The two support rollers are fixed with respect to the vertical displacement. The upper loading roller is prescribed with vertical displacement at a rate of 2 mm/min. Interaction of all specimen-to-roller contact surfaces are assumed to be frictionless.

Properties and relevant materials data for the unidirectional lamina and interface material are derived from published work on similar CFRP composite laminates [3, 5]. These properties and constants are summarized in Table 1. Subscripts 1, 2 and 3 refer to the local material axes with direction 1 along the fibers while 2 and 3 are perpendicular to the fiber orientation in the unidirectional ply.

The effect of mesh density on the predicted flexural response of the CFRP composite is examined. An initial mesh with 31200 solid elements for the laminates and 15540 cohesive elements for the interfaces (a total of 46740 elements) is used as a reference coarse element mesh with mesh density defined as $\rho_{\text{ref}} = 1.0$. This reference finite element mesh is illustrated in Fig. 3. Subsequent mesh refinements with higher relative mesh density ratio of $r = \rho/\rho_{\text{ref}} = 1.6, 2.5$ and 5.0 are employed. The resulting load–deflection responses of the CFRP composite specimen are compared in Fig. 4. The initial slope represents elastic flexural stiffness of undamaged specimen. Results indicate that a coarse mesh predicted higher starting load level to initiate delamination. The slope of the curve reduces to

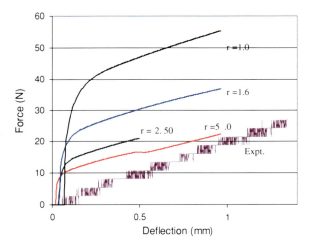

Fig. 4 Load–deflection responses of the model with different mesh density ratios, r

the measured value as the mesh density of the model increases. It is worth mentioning that the resolution of the 5-kN load cell used during the test contributed to the poor measured response of the specimen at low load level in the range of 0.02 kN. Subsequent analysis is performed using the sample geometry with the final refined mesh (r = 5.0).

5 Results and Discussion

Results of FE simulation of CFRP composite specimen under flexural loading is presented and discussed in terms of load–deflection response, initiation and progression of interface delamination, and damage dissipation energy as function of the loading.

5.1 Load–deflection Response

The predicted load–deflection curve of the CFRP composite sample under three-point bend test is compared with measured values as shown in Fig. 5. The flexural behavior of an elastic, non-damage composite model with perfectly bonded interfaces is also included for comparison. The calculated transverse load (and hence the flexural moment) is in excess of 500 N when the prescribed mid-span deflection reached 18 mm. The corresponding measured load of 248 N indicates the occurrence of severe damage in the composite sample. The measured flexural stiffness of the composite, as represented by the slope of curve is approximately 46.7% lower than that predicted by the non-damage model. This significant loss of stiffness is primarily due to interface delamination. Such flexural stiffness

Fig. 5 Comparison of predicted and measured load–deflection responses of the CFRP composite

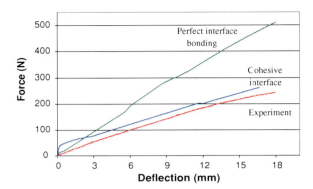

degradation also suggests the corresponding loss of lateral load-carrying capability of the CFRP composite panel following delamination damage. However, the application of in-plane tensile pre-stressing of the composite panel may results in different lateral load-carrying characteristics [9].

The predicted flexural stiffness of the CFRP composite, modeled with delaminating interfaces is similar to the measured stiffness. Nonlinear load–deflection response of the CFRP composite sample during the flexural loading is adequately reproduced. This similar slope indicates that interface delamination is the dominant damage mechanism for the given loading. A noticeable initial stiffness is reflective of elastic behavior of the laminates assumed in the model. The initiation of delamination is indicated by sharp change in slope of the curve. A constant overestimation of the loading pin reaction force magnitude is likely due to the assumption of elastic laminates behavior and partly to the mesh density effect, as discussed above.

5.2 Interface Shear Stress Distribution

Interface shear stress developing during flexural loading of the CFRP composite specimen will dictate the onset of interface delamination since the normal stress component is expected to be insignificant. Typical variations of shear stress in each interface across the thickness of the composite specimen at the center mid-span location (under the loading roller) and edge location are illustrated in Fig. 6. The equivalent interface shear stress magnitude is predicted using the non-damaging interface model and corresponds to simulated flexural test duration of 250 s. The onset of interface damage is first predicted to occur at 253.1 s in the 2nd interface (Fig. 7). At the center mid-span location (top figure) the shear stress is higher and more uniform in middle ply interfaces. Where maximum flexural stress is expected to occur near the outer surface of the specimen, the corresponding shear stress is minimal. At the free edge of the specimen, the variation of interface shear stress across the thickness is more drastic, as illustrated in the bottom figure.

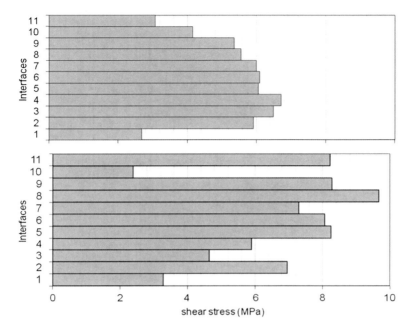

Fig. 6 Variations of interface shear stress across the thickness of the specimen at center mid-span (*top*) and edge (*bottom*) locations

The higher magnitude of stress favors initiation of edge delamination for selected interfaces. It is worth noting that interface shear stress at this edge location of the specimen is generally higher for interfaces experiencing tensile bending stress.

5.3 Progression of Interface Delamination

The predicted sequence for the onset of delamination in each interface during the three-point bend test of the specimen is presented in Fig. 7 in terms of test time duration to initiate delamination. Interface delamination initiated either in the center mid-span region or along the edges of the specimen. It is first predicted in the second interface (counted from the top) between −45/45 plies. The last interface to reach initial delamination is in the 10th interface between −45/45 plies. It is noted that delamination begins in the central region for interfaces in the upper 64% of specimen thickness where flexural stress in the laminates is primarily compressive. In addition, the initiation of delamination events in these interfaces occurs over short time duration of 160 s. Edge delamination initiated for the remaining lower interfaces where laminates are under tensile flexural stress. Fiber orientation in adjacent laminates and loading type dictate the starting location of the delamination for particular interface.

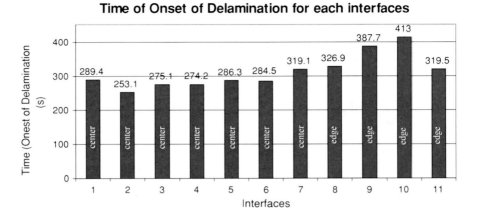

Fig. 7 Simulated test time to reach the onset of delamination in each interface

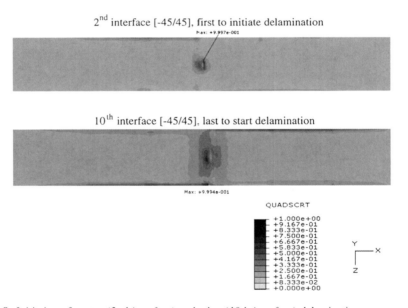

Fig. 8 Initiation of center (*2nd interface*) and edge (*10th interface*) delamination

Figure 8 shows the predicted damage state in the interfaces corresponding to the onset of center and edge delamination in the 2nd and 10th interface, respectively. Both interfaces lie between −45/45 plies. Anti-symmetric distribution of the calculated delamination region is noted and derived from the corresponding anti-symmetric ply sequence in the CFRP composite. Following delamination of

Evolution Characteristics of Delamination Damage

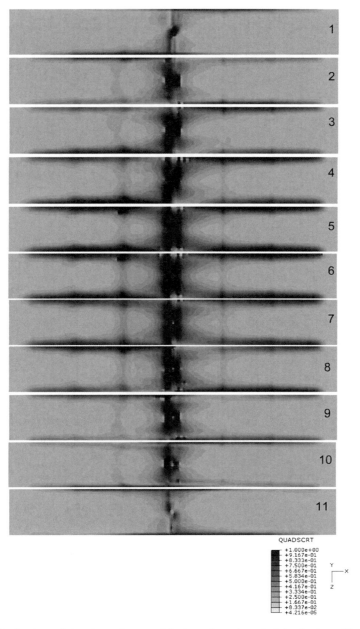

Fig. 9 Distribution of interface damage (*shades of grey*) and delamination (*black area*) corresponding to the end-of-test condition

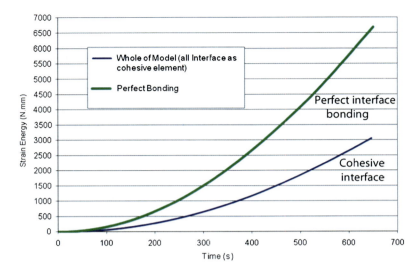

Fig. 10 Evolution of accumulated strain energy in the CFRP composite specimen as predicted by the different interface conditions

an interface material point, the corresponding shear and normal tractions at the point diminished.

The distribution of interface damage (shades of grey) and delamination (black region with the interface stress state satisfying Eq. 2 in each interface corresponding to the end-of-test condition for the CFRP composite sample is shown in Fig. 9. Both center and edge delaminations were predicted in each interface at this high flexural load. However, delamination is confined to the central region of the interface due to stress localization by the loading pin. Edge delamination is limited to a shallow depth into the interface plane. Although delamination is initiated first in the 2nd interface, the progression of delamination is extensive in middle interfaces. Anti-symmetric distribution of delaminated regions in each interface is obvious, as derived from the anti-symmetric sequence of plies lay-ups. Results also indicated that limited progression of delaminated region is sufficient to result in substantial loss of flexural stiffness of the composite.

5.4 Damage Dissipation Energy

The evolution of accumulated strain energy in the CFRP composite specimen during flexural loading is compared in Fig. 10 for model with non-damaging and damaging interfaces. Results indicate that the assumption of perfect interface bonding overestimated the strain energy by a factor of 2.2 compared to the model with cohesive damage interface. Non-linear evolution of the strain energy as

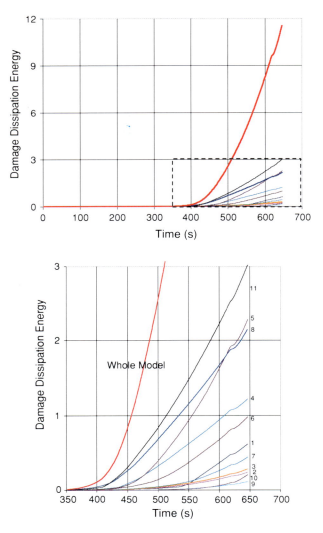

Fig. 11 Evolution of damage dissipation energy (N.mm) for interface delamination in CFRP composite. Interface number is as indicated for each curve

predicted from equivalent elastic, orthotropic lamina with non-damaging interfaces is likely derived from mechanics of laminated composite with anti-symmetric plies lay-ups. Since only interface delamination damage is being simulated in this study, the significant difference in accumulated strain energy is attributed to significant loss of stiffness of the CFRP composite specimen following interface delamination, as illustrated earlier in Fig. 5.

The progression of delamination in all interfaces is presented in Fig. 11 in terms of the evolution of damage dissipation energy for delamination. Results show that delamination failure process in CFRP composite can be described by exponential dissipation of fracture energy under monotonic three-point bend loading. Once delamination started, subsequent fast progression of failure is due to the limited toughness of the interfaces. The predicted progression rate of delamination in each interface does not follow the order of first initiation. The amount of dissipated energy for delamination in the 11th interface is five times greater than that for the 1st interface. However, the extent of delamination in the 11th interface is smaller and limited primarily to edge delamination (Fig. 9). Thus, it can be inferred that dissipation energy for edge delamination is greater than that for internal center delamination. It is worth noting that the 7th interface (between 0/45 plies) and 11th interface (between 45/−45 plies) with center and edge delamination, respectively started to delaminate at the same test time, as illustrated in Fig. 7. The subsequent rate of damage energy dissipation is higher for the edge delamination in the 11th interface. However, interface delamination damage region is smaller in the 11th interface when compared to the 7th interface at the end of the test.

6 Conclusions

The mechanics of interface delamination in CFRP composite under three-point flexural loading was examined using finite element method. Results show that:

- Interface delamination accounts for up to 46.7% reduction in flexural stiffness from the undamaged state.
- Delamination initiated at the center mid-span region for interfaces in the compressive laminates while edge delamination started in interfaces with tensile flexural stress in the laminates.
- Anti-symmetric distribution of delaminated region is derived from the corresponding anti-symmetric ply sequence in the CFRP composite.
- Delamination failure process in CFRP composite can be described by exponential fracture energy dissipation. The dissipation energy for edge delamination is greater than that for internal center delamination.

Acknowledgment CFRP laminated composite panels used in this study was fabricated at the Institute of Automotive and Transportation (ISAT), University of Burgundy, Nevers, Cedex, France. This work is an on-going collaborative research between ISAT and the Center for Composites, Universiti Teknologi Malaysia.

References

1. Benzeggagh, M.L., Kenane, M.: Measurement of mixed-mode delamination fracture toughness of unidirectional glass/epoxy composites with mixed-mode bending apparatus. Compos. Sci. Technol. **49**, 439–449 (1996)

2. Camanho, P.P., Matthews, F.L.: Delamination onset prediction in mechanically fastened joints in composite laminates. J. Compos. Mater. **33**, 906–927 (1999)
3. Campilho, R.G., Moura, M.F., Domingues, J.S.: Using a cohesive damage model to predict the tensile behavior of CFRP single-strap repairs. Int. J. Solids Struct. **45**, 1497–1512 (2007)
4. Davila, C.G., Camanho, P.P., Moura, M.F.: Mixed-mode decohesion elements for analyses of progressive delamination. Proc. 42nd AIAA/ASME/ASCE/AHS/ASC Structures, structural dynamics and materials conf. Seattle, WA (2001)
5. Davila, C.G., Camanho, P.P.: Analysis of the effects of residual strains and defects on skin/stiffener debonding using decohesion elements. Proc. 44th AIAA/ASME/ASCE/AHS structures, structural dynamics, and materials conf. Norfolk, VA (2003)
6. Krueger, R., Cvitkovich, M.K., O'Brien, T.K., Minguet, P.J.: Testing and analysis of composite skin/stringer debonding under multi-axial loading. J. Compos. Mater. **34**, 1263–1300 (2000)
7. Li, J., Sen, J.K.: Analysis of frame-to-skin joint pull-off tests and prediction of the delamination failure. Proc. 42nd AIAA/ASME/ASCE/AHS/ASC structures, structural dynamics and materials conf. Seattle, WA (2000)
8. Reeder, J. R.: An Evaluation of mixed-mode delamination failure criteria. NASA/TM-104210, VA (1992)
9. Sicot, O., Rousseau, J., Hearn, D.: Influence of stacking sequences on impact damage of prestresses isotropic composite laminates. 16th Int Conf composite materials, Japan (2007)
10. Turon, A., Camanho, P. P., Costa, J., Davila, C. G. An interface damage model for the simulation of delamination under variable-mode ratio in composite materials, NASA/TM-2004-213277, VA (2004)

Indentation of Sandwich Beams with Functionally Graded Skins and Transversely Flexible Core

Y. Mohammadi and S. M. R. Khalili

Abstract Improved high-order sandwich beam theory is used to model the local deformation under the central indenter for sandwich beams with Aluminum/ Alumina FG skins loaded under three-point bending. First shear deformation theory (FSDT) is used for the FG skins while three-dimensional elasticity is used for the flexible core. By using the model to consider the way in which different wavelengths of sinusoidal pressure loading on the top FG skin are transmitted to the core and to the bottom FG skin, two spreading length scales λ_t and λ_b are introduced and calculated. λ_t and λ_b, which are two functions of the beam material and geometric properties, characterize the length over which a load on the top surface of a beam is spread out by the skins and the core. When semi-wavelength is greater than λ_t (or λ_b), the contact load at the top FG skin is transmitted relatively unchanged to the core (or to the bottom FG skin). Conversely, when L/m $<$ λ_t (or λ_b), the applied load is spread out by the top FG skin (or by the top FG skin and the core) over a length of the order of λ_t (or λ_b). Reasonable agreement is found between theoretical predictions of the displacement field under the indentation loading and FEM results of ANSYS using a sandwich beam with functionally graded skins and transversely flexible core.

Y. Mohammadi (✉)
Faculty of Industrial and Mechanical Engineering,
Islamic Azad University, Ghazvin Branch, Ghazvin, Iran
e-mail: u.mohammadi@gmail.com

S. M. R. Khalili
Faculty of Mechanical Engineering,
Center of Excellence for Research in Advanced Materials and structures,
K.N.Toosi University of Technology, Tehran, Iran

S. M. R. Khalili
Faculty of Engineering, Kingston University of Technology,
Tehran, Iran

Keywords Sandwich beam · Indentation · First order shear deformation theory · Functionally graded skins · Flexible core

1 Introduction

Sandwich beams with vertical flexible core are more susceptible to premature failure than beams with incompressible core [1]. A typical advanced construction of a sandwich beam consists of two FG skins, not necessarily identical, bonded to the compressible core through adhesive layers. The separation of FG skins by a lightweight core increases the bending rigidity of the beam at expenses of small weight. Also, the functionally graded materials (FGMs) are multi-functional materials which contain spatial variations in composition and microstructure for the specific purpose of controlling variations in thermal, structural, or functional properties. These materials are presently in the forefront of material research receiving worldwide attention [2]. Such materials have a broad range of applications including for example, biomechanical, automotive, aerospace, mechanical, civil, nuclear, and naval engineering. FGMs are microscopically inhomogeneous composites usually made from a mixture of metals and ceramics. The considerable advantages offered by FGMs over conventional materials and the need of overcoming the technical challenges involving high temperature environments have prompted an increased use of sandwich structures, and incorporation in their construction are the FGMs as face sheets [2, 3].

Analysis of sandwich beams are presented fundamentally by Allen [4] and Plantema [5], assuming that the core is incompressible in the out-of-plane direction. These models assumed that the skins have only bending rigidity, while the core has only shear rigidity. This approach is good for sandwich structures with incompressible cores. To model the local effects at the load points for non-metallic honeycomb sandwich panels with low transverse flexibility, compressibility of the core in the vicinity of the applied loads must be included. The non-planar deformed cross-section of the sandwich beam, observed experimentally, suggested the need for a model which allows nonlinear variations of in-plane and vertical displacement field through the core [6]. Frostig and Baruch [7, 8] used variational principles to develop a high-order sandwich panel theory, which includes the transverse flexibility of the core. In contrast, the simple beam theory where the core in-plane displacements are assumed to vary in a linear way through the depth, and the out-of-plane displacements are assumed to be constant. The accuracy of Frostig's formulation as applied to indentation of sandwich panels had already been verified by Petras and Sutcliffe [6]. An improved high-order sandwich plate theory (IHSAPT) based on the First order shear deformation displacement field carried out on the Frostig's high-order sandwich panel theory by Khalili et al. [9] and Malekzadeh et al. [10]. Indentation failure of sandwich structures can arise from a number of causes including failure at the loading points or joints or due to impact damage [6].

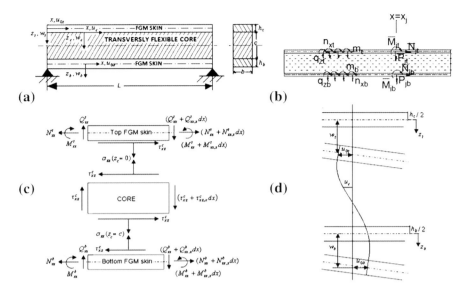

Fig. 1 Geometry, loads, and internal resultants and deformations of differential segment: **a** Geometry; **b** Loads and total internal forces; **c** Internal forces on skins and core; **d** Deformations [6, 7]

The aim of this paper is to understand the indentation behavior of sandwich panels with FG skins by examining the roles of geometrical and mechanical properties of the core and FG skins on the localized behavior under an indentation loading. IHSAPT is used in this paper to analyze the behavior of sandwich beams with FG skins subjected to three-point bending and to examine the details at the load points. Two spreading length scales that characterize the flexibility of a sandwich beam under indentation loading is also calculated. The accuracy of IHSAPT in modeling localized effects is verified by ANSYS FE code.

2 Formulation

The IHSAPT introduced by Khalili and Malekzadeh [9] is used in this paper. This high-order analysis is based on variational principles. Consider a sandwich beam of span, L and unit width, consisting of a core with thickness, c, Young's and shear modulus, E_C and G_C, respectively, and two skins with the thicknesses of h_t and h_b for the top and bottom skins, respectively, Young's modulus E_f and Poisson's ratio v_f, as depicted in Fig. 1 [6, 7].

A distributed load, q_t is applied to the top skin. The model is two-dimensional, so that variations across the width are neglected. The displacement and stress fields of the core are expressed in terms of the following seven unknowns: the in-plane deformations, u_{0t} and u_{0b} in the x direction at the mid-plane of the top and bottom skin, respectively; their corresponding vertical displacements, w_t and w_b; the rotation components of the transverse normal about the y-axes ψ_x^t and ψ_x^b of the mid-plane of the top and bottom skin, respectively, and the out of plane shear stress τ_{xz}^c in the core. The relevant geometric parameters and the notation used for stresses and displacements are given in Fig. 1.

The governing equations and boundary conditions are derived by the variational principles imposed on the total potential energy as follows:

$$\delta U_e + \delta V_e = 0 \quad (1)$$

where U_e and V_e are internal potential and external energies, respectively; and δ denotes the variation operator.

The first variation of the internal potential energy in terms of stresses and strains reads as

$$\delta U_e = \int_{V_t} \left(\sigma_{xx}^t \delta\varepsilon_{xx}^t + \tau_{xz}^t \delta\gamma_{xz}^t \right) dv + \int_{V_b} \left(\sigma_{xx}^b \delta\varepsilon_{xx}^b + \tau_{xz}^b \delta\gamma_{xz}^b \right) dv$$
$$+ \int_{V_c} \left(\sigma_{xx}^c \delta\varepsilon_{xx}^c + \tau_{xz}^c \delta\gamma_{xz}^c \right) dv \quad (2)$$

where σ_{xx}^j and ε_{xx}^j (j = t, b) are the normal stress and strain in x-direction of the upper and the lower skins, τ_{xz}^j and γ_{xz}^j (j = t, b) are the vertical shear stress and strain in skins; superscripts t and b correspond to the upper and the lower skins, respectively; σ_{zz}^c and ε_{zz}^c are the normal stress and strain in the vertical direction of the core; τ_{xz}^c and γ_{xz}^c are the vertical shear stress and shear strain in the core; v_t, v_b and v_c are the volumes of the upper and lower skins and the core, respectively. The variation of the external energy is equal to:

$$\delta V_e = -\int_0^L \left(n_{xt}\delta u_{0t} + n_{xb}\delta u_{0b} + \left(q_t + \sum_{i=1}^{NP} \frac{P_{it}}{l_{pit}} \delta_d(x - x_i) \right) \delta w_t \right.$$
$$\left. + \left(q_b + \sum_{i=1}^{NP} \frac{P_{ib}}{l_{pib}} \delta_d(x - x_i) \right) \delta w_b \right) dx \quad (3)$$

where q_t and q_b are the vertical distributed static or dynamic loads exerted on the upper and lower skins of the beam, respectively; P_{it} and P_{ib} are the vertical external concentrated loads distributed on a rectangular region: l_{pit} by one at the upper skin and l_{pib} by one at the lower skin located at $x = x_i$; NP denotes the number of concentrated loads; $\delta_d(x-x_i)$ is the Delta of Dirac function at the location of the load; n_{xj} (j = t, b) are the in-plane external loads in the longitudinal direction of the upper and lower skins; w_j and u_{0j} are the vertical deflection and in-plane displacement in x-direction of the mid-plane of each skin (j = t, b).

Indentation of Sandwich Beams

Geometry and sign convention for stresses, displacements, and loads are as appeared in Fig. 1.

Considering small deformations and rotations, the kinematic relations for the skins, based on FSDT are as follows:

$$u_j(x,z) = u_{0j}(x) + z_j \psi_x^j(x) \tag{4}$$

$$w_j(x,z) = w_{0j}(x) \tag{5}$$

where $u_{0j}(x)$, ($j = t, b$) is the in-plane deformation in the x-direction of the mid-plane of each skin panel; $w_{0j}(x)$ is the transverse deflection of each skin ($j = t, b$); ψ_x^j ($j = t, b$) the rotation components of the transverse normal about the y-axes of the mid-plane of the upper and lower skins, respectively. z_j ($j = t, b$) is the vertical coordinate of each skin and is measured downward from the mid-plane of each skin (see Fig. 1a).

The kinematic relations used, assuming small deformations, take the following form for the panel skins:

$$\varepsilon_{xx}^j(x,z) = u_{0j,x}(x) + z_j \psi_{x,x}^j(x) \tag{6}$$

$$\gamma_{xz}^j(x) = w_{0j,x}(x) + \psi_x^j(x) \tag{7}$$

Where $()_{,i}$ denotes a partial derivative with respect to i. The kinematic relations for the core

$$\varepsilon_{zz}^c(x, z_c) = w_{c,z_c}(x, z_c) \tag{8}$$

$$\gamma_{xz}^c(x, z_c) = u_{c,z_c}(x, z_c) + w_{c,x}(x, z_c) \tag{9}$$

where $u_c(x, z_c)$ and $w_c(x, z_c)$ are the in-plane displacement in the x-direction and the vertical deflection of the core, respectively, and z_c is the vertical coordinate of the core, measured downward from the upper skin–core interface (see Fig. 1a).

The compatibility conditions, assuming perfect bonding between the core and the skins, at the upper and the lower skin–core interface, ($z_c = 0, c$), read as

$$u_c(z_c = 0) = u_{0t} + \frac{h_t}{2} \psi_x^t \tag{10}$$

$$u_c(z_c = c) = u_{0b} - \frac{h_b}{2} \psi_x^b \tag{11}$$

$$w_c(z_c = 0) = w_{0t} \tag{12}$$

$$w_c(z_c = c) = w_{0b} \tag{13}$$

where h_j, ($j = t, b$) and c are the thickness of the upper and lower skins and the height of the core, respectively, (Fig. 1a).

Using the total potential energy principle (Eqs. 1–3), kinematic relations (Eqs. 6–9) and stress resultants, defined in Fig. 1c, it is possible to obtain the equations of equilibrium and the appropriate boundary conditions.

$$N^t_{xx,x} + \tau^c_{xz}(z_c = 0) + n_{xt} = 0 \tag{14}$$

$$N^b_{xx,x} - \tau^c_{xz}(z_c = c) + n_{xb} = 0 \tag{15}$$

$$Q^t_{x,x} + \sigma^c_{zz}(z_c = 0) + q_t = 0 \tag{16}$$

$$Q^b_{x,x} - \sigma^c_{zz}(z_c = c) + q_b = 0 \tag{17}$$

$$M^t_{xx,x} - Q^t_x + \tau^c_{xz}(z_c = 0)\frac{h_t}{2} = 0 \tag{18}$$

$$M^b_{xx,x} - Q^b_x + \tau^c_{xz}(z_c = c)\frac{h_b}{2} = 0 \tag{19}$$

$$\tau^c_{xz,z_c} = 0 \tag{20}$$

$$\sigma^c_{zz,z_c} + \tau^c_{xz,x} = 0 \tag{21}$$

where N^j_{xx}, M^j_{xx}, $(j = t, b)$ are stress resultants and moment resultants of the upper and lower skins; Q^j_x, $(j = t, b)$ is the distributed shear force per unit length of the edge in the x-direction of each skin. Using the last two equations of the equilibrium equations (Eqs. 20–21), constitutive law of the core material and the continuity conditions for the top and bottom skins, the analytical relations for the normal stress as well as the horizontal and vertical displacements in the core were derived by Khalili and Malekzadeh [9].

For stress resultants, moment resultants, and distributed shear force per unit length of each FG skin, constitutive equations are as follows:

$$N^j_{xx} = A^j_{11}\varepsilon^{j0}_{xx} + B^j_{11}\kappa^j_{xx} \tag{22}$$

$$M^j_{xx} = B^j_{11}\varepsilon^{j0}_{xx} + D^j_{11}\kappa^j_{xx} \quad (j = t, b) \tag{23}$$

$$Q^j_x = A^j_{55}\gamma^j_{xz} \tag{24}$$

where $\kappa^j_{xx} = \psi^j_{x,x}$, $(j = t, b)$ is the curvature in the x-direction; ε^{j0}_{xx}, $(j = t, b)$ is the mid-plane normal strain component; and γ^j_{xz}, $(j = t, b)$ is the out-of-plane shear strain component. The terms, A^j_{11}, B^j_{11}, D^j_{11} and A^j_{55}, $(j = t, b)$ for upper and lower functionally graded skins can be introduced as follows [11]:

Power-law function of FGM (P-FGM):

$$\left\{\begin{array}{c}A_{11}^t\\B_{11}^t\\D_{11}^t\end{array}\right\} = \int_{-\frac{h_t}{2}}^{\frac{h_t}{2}} \frac{(E_2 - E_1)\left(\frac{z_t + h_t/2}{h_t}\right)^p + E_1}{1 - \left((\nu_2 - \nu_1)\left(\frac{z_t + h_t/2}{h_t}\right)^p + \nu_1\right)^2} \left\{\begin{array}{c}1\\z_t\\z_t^2\end{array}\right\} dz_t \quad (25)$$

$$\left\{\begin{array}{c}A_{11}^b\\B_{11}^b\\D_{11}^b\end{array}\right\} = \int_{-\frac{h_b}{2}}^{\frac{h_b}{2}} \frac{(E_3 - E_4)\left(\frac{-z_b + h_b/2}{h_b}\right)^p + E_4}{1 - \left((\nu_3 - \nu_4)\left(\frac{-z_b + h_b/2}{h_b}\right)^p + \nu_4\right)^2} \left\{\begin{array}{c}1\\z_b\\z_b^2\end{array}\right\} dz_b \quad (26)$$

$$A_{55}^t = \int_{-\frac{h_t}{2}}^{\frac{h_t}{2}} \frac{(E_2 - E_1)\left(\frac{z_t + h_t/2}{h_t}\right)^p + E_1}{2\left(1 + (\nu_2 - \nu_1)\left(\frac{z_t + h_t/2}{h_t}\right)^p + \nu_1\right)} dz_t \quad (27)$$

$$A_{55}^b = \int_{-\frac{h_b}{2}}^{\frac{h_b}{2}} \frac{(E_3 - E_4)\left(\frac{-z_b + h_b/2}{h_b}\right)^p + E_4}{2\left(1 + (\nu_3 - \nu_4)\left(\frac{-z_b + h_b/2}{h_b}\right)^p + \nu_4\right)} dz_b \quad (28)$$

Exponential function of FGM (E-FGM):

$$\left\{\begin{array}{c}A_{11}^t\\B_{11}^t\\D_{11}^t\end{array}\right\} = \int_{-\frac{h_t}{2}}^{\frac{h_t}{2}} \frac{E_2 \exp\left(\log\frac{E_2}{E_1}(z_t/h_t - 1/2)\right)}{1 - \left(\nu_2 \exp\left(\log\frac{\nu_2}{\nu_1}(z_t/h_t - 1/2)\right)\right)^2} \left\{\begin{array}{c}1\\z_t\\z_t^2\end{array}\right\} dz_t \quad (29)$$

$$\left\{\begin{array}{c}A_{11}^b\\B_{11}^b\\D_{11}^b\end{array}\right\} = \int_{-\frac{h_b}{2}}^{\frac{h_b}{2}} \frac{E_4 \exp\left(\log\frac{E_4}{E_3}(z_b/h_b - 1/2)\right)}{1 - \left(\nu_4 \exp\left(\log\frac{\nu_4}{\nu_3}(z_b/h_b - 1/2)\right)\right)^2} \left\{\begin{array}{c}1\\z_b\\z_b^2\end{array}\right\} dz_b \quad (30)$$

$$A_{55}^t = \int_{-\frac{h_t}{2}}^{\frac{h_t}{2}} \frac{E_2 \exp\left(\log\frac{E_2}{E_1}(z_t/h_t - 1/2)\right)}{2\left(1 + \nu_2 \exp\left(\log\frac{\nu_2}{\nu_1}(z_t/h_t - 1/2)\right)\right)} dz_t \quad (31)$$

$$A_{55}^b = \int_{-\frac{h_b}{2}}^{\frac{h_b}{2}} \frac{E_4 \exp\left(\log\frac{E_4}{E_3}(z_b/h_b - 1/2)\right)}{2\left(1 + \nu_4 \exp\left(\log\frac{\nu_4}{\nu_3}(z_b/h_b - 1/2)\right)\right)} dz_b \quad (32)$$

Sigmoid distribution of FGM (S-FGM)

Fig. 2 Structure of sandwich beam with FG skins and mechanical properties notations of upper and lower surfaces of both top and bottom FG skins

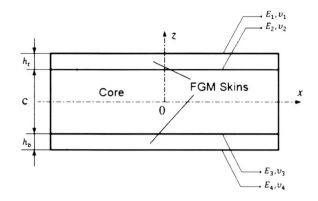

$$\left\{\begin{array}{c} A_{11}^t \\ B_{11}^t \\ D_{11}^t \end{array}\right\} = \int_0^{\frac{h_t}{2}} \frac{E_2 + \frac{(E_1-E_2)}{2}\left(\frac{h_t/2-z_t}{h_t/2}\right)^p}{1 - \left(v_2 + \frac{(v_1-v_2)}{2}\left(\frac{h_t/2-z_t}{h_t/2}\right)^p\right)^2} \left\{\begin{array}{c} 1 \\ z_t \\ z_t^2 \end{array}\right\} dz_t$$

$$+ \int_{-\frac{h_t}{2}}^0 \frac{\frac{(E_2-E_1)}{2}\left(\frac{z_t+h_t/2}{h_t/2}\right)^p + E_1}{1 - \left(\frac{(v_2-v_1)}{2}\left(\frac{z_t+h_t/2}{h_t/2}\right)^p + v_1\right)^2} \left\{\begin{array}{c} 1 \\ z_t \\ z_t^2 \end{array}\right\} dz_t \qquad (33)$$

$$\left\{\begin{array}{c} A_{11}^b \\ B_{11}^b \\ D_{11}^b \end{array}\right\} = \int_0^{\frac{h_b}{2}} \frac{E_4 + \frac{(E_3-E_4)}{2}\left(\frac{h_b/2-z_b}{h_b/2}\right)^p}{1 - \left(v_4 + \frac{(v_3-v_4)}{2}\left(\frac{h_b/2-z_b}{h_b/2}\right)^p\right)^2} \left\{\begin{array}{c} 1 \\ z_b \\ z_b^2 \end{array}\right\} dz_b$$

$$+ \int_{-\frac{h_b}{2}}^0 \frac{E_3 + \frac{(E_4-E_3)}{2}\left(\frac{z_b+h_b/2}{h_b/2}\right)^p}{1 - \left(v_3 + \frac{(v_4-v_3)}{2}\left(\frac{z_b+h_b/2}{h_b/2}\right)^p\right)^2} \left\{\begin{array}{c} 1 \\ z_b \\ z_b^2 \end{array}\right\} dz_b \qquad (34)$$

$$A_{55}^t = \int_0^{\frac{h_t}{2}} \frac{E_2 + \frac{(E_1-E_2)}{2}\left(\frac{h_t/2-z_t}{h_t/2}\right)^p}{2\left(1 + v_2 + \frac{(v_1-v_2)}{2}\left(\frac{h_t/2-z_t}{h_t/2}\right)^p\right)} \left\{\begin{array}{c} 1 \\ z_t \\ z_t^2 \end{array}\right\} dz_t$$

$$+ \int_{-\frac{h_t}{2}}^0 \frac{\frac{(E_2-E_1)}{2}\left(\frac{z_t+h_t/2}{h_t/2}\right)^p + E_1}{2\left(1 + \frac{(v_2-v_1)}{2}\left(\frac{z_t+h_t/2}{h_t/2}\right)^p + v_1\right)} \left\{\begin{array}{c} 1 \\ z_t \\ z_t^2 \end{array}\right\} dz_t \qquad (35)$$

$$A_{55}^b = \int_0^{\frac{h_b}{2}} \frac{E_4 + \frac{(E_3-E_4)}{2}\left(\frac{h_b/2-z_b}{h_b/2}\right)^p}{2\left(1 + v_4 + \frac{(v_3-v_4)}{2}\left(\frac{h_b/2-z_b}{h_b/2}\right)^p\right)} \begin{Bmatrix} 1 \\ z_b \\ z_b^2 \end{Bmatrix} dz_b$$
$$+ \int_{-\frac{h_b}{2}}^0 \frac{E_3 + \frac{(E_4-E_3)}{2}\left(\frac{z_b+h_b/2}{h_b/2}\right)^p}{2\left(1 + v_3 + \frac{(v_4-v_3)}{2}\left(\frac{z_b+h_b/2}{h_b/2}\right)^p\right)} \begin{Bmatrix} 1 \\ z_b \\ z_b^2 \end{Bmatrix} dz_b \quad (36)$$

where E_1 and E_2 are the Young's modulus of the lowest $(z_t = h_t/2)$ and top surfaces $(z_t = -h_t/2)$ of the upper FG skin, respectively, E_3 and E_4 are the Young's modulus of the top $(z_b = -h_b/2)$ and lowest surfaces $(z_b = h_b/2)$ of the lower FG skin, respectively; v_i, $(i = 1, 2, 3, 4)$ are the related Poisson's ratios (see Fig. 2); and p is a positive constant exponent of FGM.

3 Governing Equations

Finally, the governing equations are formulated in terms of the following seven unknowns: the in-plane deformations in the x-direction of the centroid plane of skin beams, u_{0t} and u_{0b}; the transverse deformations of the various skins, w_t and w_b; the rotation components of the transverse normal about the y-axes of the mid-plane of each skin, ψ_x^t and ψ_x^b; and the transverse shear stress in the core, τ_{xz}^c. Hence, six equations are determined through substitution of the constitutive relations in Eqs. 22–24, in the field equations in Eqs. 14–19. The additional equation is derived using the displacement distribution in the x-direction of the core and the compatibility requirements at the lower skin-core interface in Eq. 11. Hence the governing equations read as

$$A_{11}^t u_{0t,xx} + B_{11}^t \psi_{x,xx}^t + \tau_{xz}^c + n_{xt} = 0 \quad (37)$$

$$A_{11}^b u_{0b,xx} + B_{11}^b \psi_{x,xx}^b - \tau_{xz}^c + n_{xb} = 0 \quad (38)$$

$$A_{55}^t k \left(w_{0t,xx} + \psi_{x,x}^t \right) + \frac{E_c}{c}(w_{0b} - w_{0t}) + \tau_{xz,x}^c \frac{c}{2} + q_t = 0 \quad (39)$$

$$A_{55}^b k \left(w_{0b,xx} + \psi_{x,x}^b \right) - \frac{E_c}{c}(w_{0b} - w_{0t}) + \tau_{xz,x}^c \frac{c}{2} + q_b = 0 \quad (40)$$

$$B_{11}^t u_{0t,xx} + D_{11}^t \psi_{x,xx}^t - A_{55}^t k \left(w_{0t,x} + \psi_x^t \right) + \tau_{xz}^c \frac{h_t}{2} = 0 \quad (41)$$

$$B_{11}^b u_{0b,xx} + D_{11}^b \psi_{x,xx}^b - A_{55}^b k \left(w_{0b,x} + \psi_x^b \right) + \tau_{xz}^c \frac{h_b}{2} = 0 \quad (42)$$

$$\frac{\tau_{xz}^c c}{G_c} - \frac{\tau_{xz,xx}^c c^3}{(12E_c)} - \frac{c}{2}(w_{0t,x} + w_{0b,x}) + \frac{1}{2}(h_t \psi_x^t + h_b \psi_x^b) + u_{0t} - u_{0b} = 0 \qquad (43)$$

where k is the shear correction ratio that according to Mindlin's assumption [12], is taken as $\pi^2/12$. The solution of the set equations can be achieved numerically for general type of boundary conditions or analytically for the particular type of boundary conditions such as simply supported.

4 Simply Supported Beam

An analytical solution exists in the case of a simply supported sandwich beam where the two skin beams are simply supported and the vertical deformation through the depth of the core at the edges of the beams is prevented. This solution consists of an infinite series of trigonometric functions in x-direction, a Fourier series in one dimension provided that the load exerted on the various skins, which may be localized distributed or fully distributed, are described in terms of trigonometric functions. The solution series reads as

$$\begin{bmatrix} u_{0j}(x) \\ w_{0j}(x) \\ \psi_x^j(x) \\ \tau_{xz}^c(x) \end{bmatrix} = \sum_{m=1}^{M} \begin{bmatrix} C_{uj}^m \cos(\alpha_m x) \\ C_{wj}^m \sin(\alpha_m x) \\ C_{\psi j}^m \cos(\alpha_m x) \\ C_{\tau}^m \cos(\alpha_m x) \end{bmatrix} \qquad (44)$$

where m is an index for the wavelength of the Fourier term and M the number of terms in the Fourier series. The Fourier coefficients, C_{uj}^m, C_{wj}^m, $C_{\psi j}^m$ and C_{τ}^m, $(j=t,b)$ are constants to be determined. For the external loads they are exerted on the top and bottom skins, so in terms of the Fourier series we set

$$q_j(x) = \sum_{m=1}^{M} C_{qj}^m \sin(\alpha_m x) \qquad (45)$$

where C_{qj}^m, $(j=t,b)$ are two constants that depend on the distribution of the external loads.

After substituting every term of the Fourier series (Eq. 44) into the governing equations (Eqs. 37–43), the problem can be expressed in matrix form as

$$[\mathcal{S}] \cdot [\mathbb{C}] = [\mathcal{F}] \qquad (46)$$

Where

$$[\mathcal{S}] = \begin{bmatrix} \mathcal{S}_{11} & 0 & 0 & 0 & \mathcal{S}_{15} & 0 & -1 \\ & \mathcal{S}_{22} & 0 & 0 & 0 & \mathcal{S}_{26} & 1 \\ & & \mathcal{S}_{33} & \mathcal{S}_{34} & \mathcal{S}_{35} & 0 & \mathcal{S}_{37} \\ & & & \mathcal{S}_{44} & 0 & \mathcal{S}_{46} & \mathcal{S}_{37} \\ & & & & \mathcal{S}_{55} & 0 & \mathcal{S}_{57} \\ & & & & & \mathcal{S}_{66} & \mathcal{S}_{67} \\ & & & & & & \mathcal{S}_{77} \end{bmatrix} \quad (47)$$

$$[\mathbb{C}] = \begin{bmatrix} C_{ut}^m \\ C_{ub}^m \\ C_{wt}^m \\ C_{wb}^m \\ C_{\psi t}^m \\ C_{\psi b}^m \\ C_{\tau}^m \end{bmatrix} \quad (48)$$

$$[\mathcal{F}] = \begin{bmatrix} 0 \\ 0 \\ C_{qt}^m \\ C_{qb}^m \\ 0 \\ 0 \\ 0 \end{bmatrix} \quad (49)$$

where $\mathcal{S}_{11} = A_{11}^t \alpha_m^2$, $\mathcal{S}_{22} = A_{11}^b \alpha_m^2$, $\mathcal{S}_{33} = kA_{55}^t \alpha_m^2 + E_c/c$, $\mathcal{S}_{44} = kA_{55}^b \alpha_m^2 + E_c/c$, $\mathcal{S}_{55} = D_{11}^t \alpha_m^2 + kA_{55}^t$, $\mathcal{S}_{66} = D_{11}^b \alpha_m^2 + kA_{55}^b$, $\mathcal{S}_{77} = -(c/G_c + c^3 \alpha_m^2/12E_c)$, $\mathcal{S}_{15} = B_{11}^t \alpha_m^2$, $\mathcal{S}_{26} = B_{11}^b \alpha_m^2$, $\mathcal{S}_{34} = -E_c/c$, $\mathcal{S}_{35} = kA_{55}^t \alpha_m$, $\mathcal{S}_{37} = \alpha_m c/2$, $\mathcal{S}_{46} = kA_{55}^b \alpha_m$, $\mathcal{S}_{57} = -h_t/2$ and $\mathcal{S}_{67} = -h_b/2$; and the matrix S is a symmetric matrix too.

Equation 46 is solved for $\mathbb{C}$ by using Matlab software, and the seven unknown Fourier coefficients: C_{ut}^m, C_{ub}^m, C_{wt}^m, C_{wb}^m, $C_{\psi t}^m$, $C_{\psi b}^m$ and C_{τ}^m are obtained.

5 The Spreading Length Scale for Indentation Loading

In this section, the improved high-order sandwich beam theory is used to investigate how the transmission of contact pressure under an indentor is to the core of sandwich beam, where indentation failure generally occurs. The behavior is assumed as elastic up to maximum load and the distribution of contact pressure is arbitrary, so that the results will be useful to model the failure of beams made with isotropic core and functionally graded skins. The spreading length scale introduced by Petras and Sutcliffe [6] for top skin, is calculated in the present study for sandwich beams with FG skins for both top, λ_t, and bottom skins, λ_b, in which is

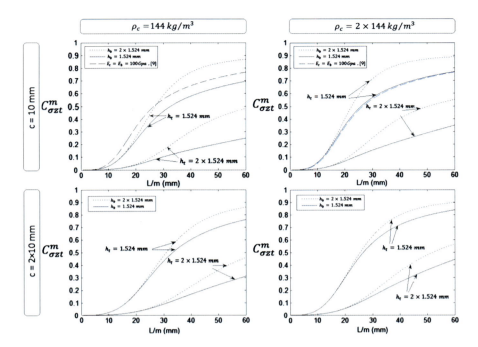

Fig. 3 Variation of dimensionless transmission coefficient of the top FG skin–core interface $C_{\sigma zt}^m$ with semi-wavelength L/m for various values of h_t, h_b, c and ρ_c. The boxed labels identify the values of ρ_c and c for each subplot

described whether the skins act in a rigid or a flexible manner. The analysis in Sect. 2 was used in which a transverse pressure on the beam is expressed in terms of a Fourier series (Eq. 45). Substituting Fourier series of w_{0t}, w_{0b} and τ_{xz} from Eq. 44 and the Fourier coefficients into the normal stress relation [13], the normal stresses in the top FG skin-core interface and the bottom FG skin–core interface are given as

$$\left\{ \begin{array}{c} \sigma_{zz}^c(z_c = 0) \\ \sigma_{zz}^c(z_c = c) \end{array} \right\} = \sum_{m=1}^{M} \left\{ \begin{array}{c} C_{\sigma zt}^m \\ C_{\sigma zb}^m \end{array} \right\} \cdot C_{qt}^m \sin(\alpha_m x) = \left\{ \begin{array}{c} C_{\sigma zt}^m \\ C_{\sigma zb}^m \end{array} \right\} \cdot q_t(x) \quad (50)$$

where $C_{\sigma zt}^m$, and $C_{\sigma zb}^m$ are the dimensionless transmission coefficients of the top and bottom FG skin–core interface, respectively, given by

$$C_{\sigma zt}^m = \frac{E_c(C_{wb}^m - C_{wt}^m)}{c} - \frac{c\alpha_m}{2} C_\tau^m \quad (51)$$

$$C_{\sigma zb}^m = \frac{E_c(C_{wb}^m - C_{wt}^m)}{c} + \frac{c\alpha_m}{2} C_\tau^m \quad (52)$$

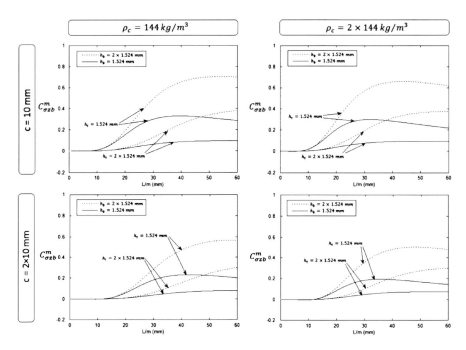

Fig. 4 Variation of dimensionless transmission coefficient of the bottom FG skin–core interface $C_{\sigma zb}^m$ with semi-wavelength L/m for various values of h_t, h_b, c and ρ_c. The boxed labels identify the values of ρ_c and c for each subplot

Table 1 The geometric and material parameters used in the parametric study

Young's modulus (GPa)								
	E_1	E_2	E_3	E_4	c(mm)	$\rho_c \left(\frac{\text{kg}}{\text{m}^3}\right)$	h_t(mm), h_b(mm)	L(mm)
Present study	380	70	70	380	10	144	1.524	60
[6]	100	100	100	100	10	144	1.524	60

Table 2 Change of spreading lengths by use of P-FGM relations for $p = 1$

	Parameter changed	λ_t(mm)	λ_b(mm)
Example1	Base sample; present study from Table 1	20.13	19.57
Example2	$c = 2 \times 10$ mm	23.81	23.39
Example3	$\rho_c = 2 \times 144\,\text{kg/m}^3$	16.89	16.41
Example4	$h_t = 2 \times 1.524$ mm	25.75	23.29
Example5	$h_b = 2 \times 1.524$ mm	22.45	23.09
Example6	$h_t = h_b = 2 \times 1.524$ mm	33.61	33.09
Example7	$L = 2 \times 60$ mm	20.06	19.57
[6]	from Table 1	17.2	–

Table 3 The values of spreading length of top skin–core interface λ_t calculated by use of all three methods, P-FGM, E-FGM, and S-FGM, for each of the examples introduced in Table 2

		$\lambda_t(mm)$						
		Exa.1	Exa.2	Exa.3	Exa.4	Exa.5	Exa.6	Exa.7
P-FGM	p = 0	15.63	18.27	13.1	21.15	17.27	26.15	15.63
	p = 0.1	17.1	20.2	14.35	22.7	19	28.61	17.1
	p = 1	20.1	23.81	16.87	25.75	22.45	33.57	20.1
	p = 5	22.1	26.37	18.6	27.75	24.87	36.97	22.1
	p = 10	22.8	27.17	19.17	28.45	25.63	38.1	22.8
E-FGM		19.26	22.81	16.15	24.93	21.47	32.2	19.26
S-FGM	p = 0	21	24.93	17.65	26.47	23.53	35.03	21
	p = 0.1	20.93	24.89	17.63	26.47	23.47	34.97	20.93
	p = 1	20.1	23.81	16.87	25.75	22.45	33.57	20.1
	p = 5	19.05	22.51	16	24.81	21.21	31.81	19.05
	p = 10	18.87	22.3	15.87	24.65	21.05	31.57	18.87

Table 4 The values of spreading length of bottom skin–core interface λ_b calculated by use of all three methods, P-FGM, E-FGM, and S-FGM, for each of the examples introduced in Table 2

		$\lambda_b(mm)$						
		Exa.1	Exa.2	Exa.3	Exa.4	Exa.5	Exa.6	Exa.7
P-FGM	p = 0	15.23	18.45	12.81	18.15	18.11	25.6	15.23
	p = 0.1	16.64	20.1	14.00	19.86	19.76	28.08	16.64
	p = 1	19.57	23.39	16.43	23.29	23.09	33.09	19.57
	p = 5	21.62	25.68	18.12	25.56	25.46	36.48	21.62
	p = 10	22.25	26.43	18.7	26.25	26.21	37.65	22.25
E-FGM		18.75	22.43	15.75	22.35	22.17	31.63	18.75
S-FGM	p = 0	20.43	24.31	17.13	24.13	24.09	34.51	20.43
	p = 0.1	20.38	24.32	17.14	24.14	24.1	34.46	20.38
	p = 1	19.57	23.39	16.43	23.29	23.09	33.09	19.57
	p = 5	18.54	22.18	15.6	22.12	21.9	31.3	18.54
	p = 10	18.37	22.01	15.43	22.01	21.73	31.07	18.37

These dimensionless coefficients depend on the geometric and material properties of the sandwich beam and on L/m, the semi-wavelength of the mth term of the Fourier series. These coefficients vary from zero for small L/m, to one for large L/m [6]. A value of zero for $C_{\sigma zt}^m$, and $C_{\sigma zb}^m$ implies no transmission of this Fourier component of applied stress through the upper FG skin, and through the core and upper FG skin, respectively, while a value of one implies total transmission, so that the parameters characterize the transparency of the skins as a function of the wavelength of the applied load.

The results of a parametric study plotted in Figs. 3 and 4, show the variation in transmission coefficients $C_{\sigma zt}^m$, and $C_{\sigma zb}^m$ with the semi-wavelength L/m of the applied sinusoidal load on the top FG skin, respectively. Table 1 shows the values of material and geometrical parameters considered in this study and the values of parameters in Ref. [6]. Each subplot in Figs. 3 and 4 contains curves for two values of top FG skin thickness h_t and bottom FG skin thickness h_b, while changing from one subplot to the next represents a change in either core density ρ_c or core thickness c. Figures 3 and 4 show that, at short wavelengths, the transmission coefficients $C_{\sigma zt}^m$, and $C_{\sigma zb}^m$ fall to zero, indicating that the contact stresses at these short wavelengths are changed significantly as they pass through the top FG skin to the core in Fig. 3 and through the top FG skin and core to the bottom FG skin in Fig. 4. The values of λ_t and λ_b in Table 2 indicate that the tendency of the skin to spread out the load (or to increasing λ_t or λ_b) increases with increasing the top FG skin thickness h_t, core thickness c and bottom FG skin thickness h_b. The effect of c is more than h_b and the effect of h_t is more than both h_b and c for change in the value of top spreading length scale, λ_t, while for change in the value of bottom FGM length scale, λ_b, the effect of h_t is a little more than h_b and the effect of c is a little more than both of them in this case. The results of Table 2 show increase in the core density, ρ_c, decreases the spreading length scales. Separate calculations show that spreading lengths did not depend on the beam length, L.

In Figs. 3 and 4, it can be seen that the inflection point in each curve can be used to identify where $C_{\sigma zt}^m$, or $C_{\sigma zb}^m$ becomes small. The semi-wavelength, L/m, at this inflection point, which we denote as the spreading length, λ_t or λ_b, characterizes the susceptibility of sandwich beams to localized effects. Contact stresses with semi-wavelengths below λ_t (or λ_b) are spread out by the top FG skin (or by the both top FG skin and core), while for longer wavelengths the applied stresses pass through the top FG skin to the core (or pass through the both top FG skin and core to the bottom FG skin) relatively unchanged.

The first row in Table 2 is a base sample, using the values of "present study" from Table 1 and named "Example 1", while subsequent rows that are named "Example 2" to "Example 7" change one of the parameters as indicated. The final row is an example of sandwich beam with non FG skins from Ref. [6] for comparison with results of sandwich beams with FG skins in this study. Table 2 shows that top skin thickness h_t is the most significant factor. Thus the thicker skins are able to spread the load better than the thin skins.

All results of Table 2 were calculated by use of power-law relations (Eqs. 25–28) and for linear distribution of properties through the thickness (p = 1), but in Tables 3 and 4 the values of spreading lengths are calculated by use of all three methods, P-FGM (Eqs. 25–28), E-FGM (Eqs. 29–32), and S-FGM (Eqs. 33–36) mentioned in Sect. 2, and for five different values of p from zero to ten. These results are listed in Tables 3 and 4 for each of the examples introduced in Table 2. The first and seventh columns of Tables 3 and 4 that show the results of Example 1 and Example 7 of Table 2, respectively, have identical results indicating that the spreading lengths do not depend on the beam length as expected. The third and ninth rows of Tables 3 and 4 also have the same results that show for linear distribution of

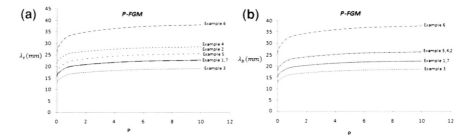

Fig. 5 The spreading length scales versus power exponent p by use of P-FGM property distribution relations, for each of the examples introduced in Table 2: **a** the spreading length scale of top skin–core interface λ_t and **b** bottom skin–core interface λ_b

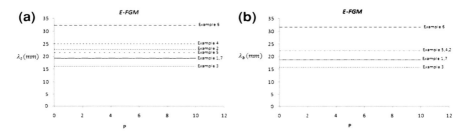

Fig. 6 The spreading length scales versus power exponent p by use of E-FGM property distribution relations, for each of the examples introduced in Table 2: **a** the spreading length scale of top skin–core interface λ_t and **b** bottom skin–core interface λ_b

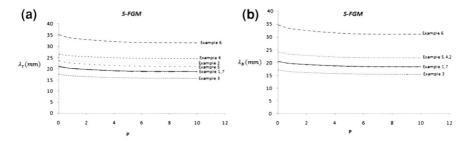

Fig. 7 The spreading length scales versus power exponent p by use of S-FGM property distribution relations, for each of the examples introduced in Table 2: **a** the spreading length scale of top skin–core interface λ_t and **b** bottom skin–core interface λ_b

properties through the thickness (p = 1) that two P-FGM and S-FGM relations illustrated in Sect. 2 have identical results as expected.

Figures 5a, 6a, and 7a plotted the results of Table 3 for each of the P-FGM, E-FGM, and S-FGM property distribution relations, respectively, versus p. and Figs. 5b, 6b, and 7b show the similar curves from the listed results of Table 4. Figures 5 and 7 show that, the achieved values of spreading lengths, λ_t and λ_b, by

Indentation of Sandwich Beams

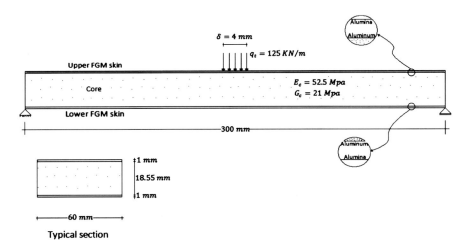

Fig. 8 Loads and geometrical and mechanical properties of typical sandwich beam with FG skins

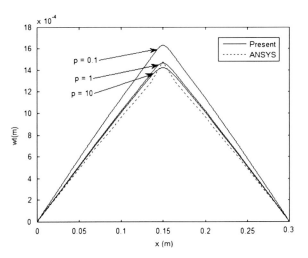

Fig. 9 Predicted and FEM results of top FG skin deflection for p = 0.1, 1 and 10

P-FGM and S-FGM relations increase and decrease, respectively, with increasing p. In additon, we can seen that the values of λ_t and λ_b calculated by P-FGM distribution relations (Eqs. 25–28) also change with p much more than the values calculated by S-FGM distribution relations (Eqs. 33–36). Also, we knew from Eqs. 29–32 that the E-FGM distribution relations did not depend on the p (see Fig. 6).

In summary, dimensionless "transmission coefficients", $C_{\sigma zt}^m$, and $C_{\sigma zb}^m$, describe how indentation loads are transmitted through the top FG skin into the core and through the top FG skin and core into the bottom FG skin, respectively. Plots of

Fig. 10 Predicted and FEM results of bottom FG skin deflection for p = 0.1, 1 and 10

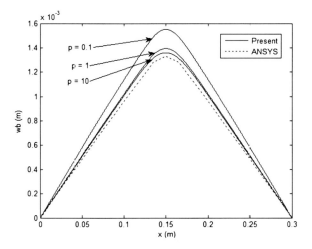

Fig. 11 The indentation value of each typical section of the sandwich beam with FG skins for three magnitudes of power exponent p

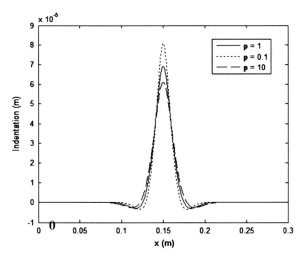

$C_{\sigma zt}^m$, and $C_{\sigma zb}^m$ as a function of load semi-wavelength, L/m, (see Figs. 3 and 11) are used to achieved two characteristic wavelengths, λ_t and λ_b, whereby below these values the contact stresses are changed significantly by the skins and core, for a given beam geometry and material combination. This analysis, which is based on elastic behavior, allows us to understand the beam behavior for a given applied pressure distribution.

Fig. 12 Predicted and FEM results of top FG skin longitudinal displacement for p = 0.1, 1 and 10

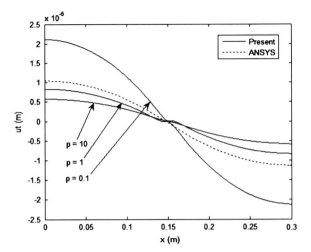

Fig. 13 Predicted and FEM results of bottom FG skin longitudinal displacement for p = 0.1, 1 and 10

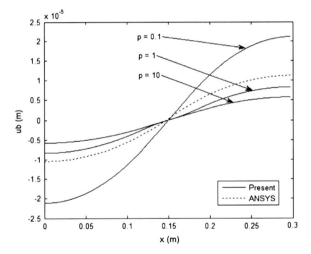

6 Verification and Numerical Results

In order to verify and validate the results for indentation of sandwich beams with FG skins, the numerical results obtained are compared with FEM results achieved from ANSYS. The present analysis results in this section are obtained by use of power-law function of FGM relations, Eqs. 25–28.

The numerical study consists of a symmetric sandwich beam with FG skins loaded by q_t at mid-span of the upper skin, where the beam is simply supported at the edges of the upper and lower skins, see Fig. 8. The skins are assumed to be

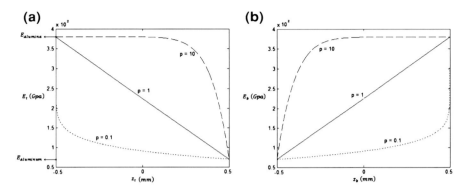

Fig. 14 Variation of the elasticity modulus through the thickness of the **a** top FG skin, and **b** bottom FG skin, according to p. by use of P-FGM property distribution

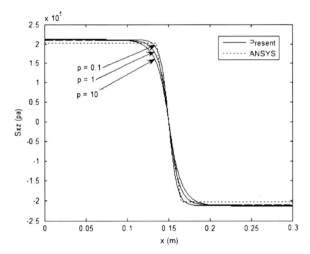

Fig. 15 Predicted and FEM results of shear stress in the core for $p = 0.1$, 1 and 10

functionally graded materials made of Aluminum/Alumina, where their modulus of elasticity are 70 and 380 GPa, respectively. The core is a lightweight closed cells foam with a density of $50\,\text{kg}/\text{m}^2$.

The numerical results include vertical and longitudinal displacements of the top and bottom FG skins, shear stress in the core, and interfacial vertical normal stresses in the top FG skin–core and bottom FG skin–core interface.

Figure 9 shows the vertical displacement of the top FG skin for the example mentioned in Fig. 8, the results of which are measured based on the power-law function of FGM relations and for three various values of exponent p. Figure 10 indicates the vertical displacement of the bottom FG skin. In both Figs. 9 and 10, the calculated results by the present analysis are verified with the FEM results

Fig. 16 Interface vertical normal stresses at core-skin interfaces, for three values of p

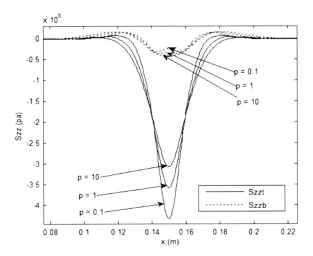

achieved from ANSYS. It can be seen that the numerical results are in good agreement with the FEM results.

Figure 11 reports the indentation value of each typical section of the sandwich beam for three types of property distribution in the FG skins, and it can be seen that the indentation effects are localized in the vicinity of the indenter as expected.

Figures 12 and 13 show the longitudinal displacements of the top and bottom FG skins, respectively, for three values of exponent p. The previous figures (Figs. 9, 10, 11, 12 and 13) show that the property distribution of FG skins by power-law relations with exponent p = 0.1 has more displacements than the two other types of property distributions, p = 1 and p = 10, and displacement results of property distribution with exponent p = 1 is a little more than similar results with exponent p = 10. As indicated in Fig. 14, modulus of elasticity for P-FGM property distribution with exponent p = 0.1 tends to the elasticity modulus of Aluminum and inverse for exponent p = 10. It can be seen that it tends to the elasticity modulus of Alumina. And because the elasticity modulus of Alumina is more than elasticity modulus of Aluminum, hence it shows bigger displacements for exponent p = 0.1 than exponent p = 10 as shown in Figs. 9, 10, 11, 12 and 13.

The results of shear stress in the core plotted in Fig. 15 for various power exponents p and compared with FEM results from ANSYS, show that the results of the present study agree very well with FEM results as expected.

Figure 16 shows the normal stresses in the top FG skin–core interface, σ_{zz}^t, together with the bottom FG skin–core interface, σ_{zz}^b. These results are plotted for three property distributions of FG skins. Normal stresses in the top FG skin–core interface, which are much greater than normal stresses in the bottom FG skin–core interface as indicated in Fig. 16, are to be expected. Because it is known from Sect. 3 that the dimensionless coefficients, $C_{\sigma zt}^m$, and $C_{\sigma zb}^m$, describe how the indentation loads are transmitted through the top skin into the core and through the top skin and the core into the bottom skin, respectively, and Figs. 3 and 4 show that the $C_{\sigma zb}^m$ is smaller than the $C_{\sigma zt}^m$, because the indentation loads, q_t, to reach to the top skin–core interface are modified by just the top FG skin, but to reach to the bottom skin–core interface, indentation loads are modified by both the core and the top FG skin. Thus, for the same reason, the peel stresses in the bottom skin–core interface are smaller than peel stresses in the top skin–core interface as shown in Fig. 16.

7 Conclusions

An improved high-order sandwich beam theory is used to analyze the indentation behavior of sandwich beams with properties typical of transversely flexible core and FG skins. The capabilities of improved high-order sandwich beam theory for sandwich beams with FG skins and transversely flexible core are verified by FEM results of the vertical and longitudinal displacements and shear stresses in the core. Although there are some quantitative differences, the results are encouraging and justify the use of the model in an analysis of indentation behavior of such beams. By considering the general loading case using a Fourier series, the effect of the applied wavelength of the contact stress on the beam behavior is investigated. Results are used to extract two spreading length scales λ_t and λ_b for a given beam material and geometric properties. This is a property of a sandwich beam, characterizing the susceptibility of the sandwich beam to indentation loads. Small values of λ_t and λ_b transmit external loads straight into the core and into the bottom FG skin, respectively, while large values of λ_t and λ_b spread the indentation load over a wider area. Spreading lengths are calculated by use of each of the P-FGM, E-FGM, and S-FGM property distribution relations of FG skins for the change in the geometrical and mechanical properties of the sandwich beam with Aluminum/Alumina FG skins. Results show that the top skin thickness h_t, core thickness c, and bottom skin thickness h_b are the most significant factors, respectively, by varying the value of λ_t. Also, by changing the value of λ_b, the most important factors are core thickness c, top skin thickness h_t and bottom skin thickness h_b, respectively. Also, the results of λ_b indicate that the effects of c, h_t and h_b in change λ_b are similar. Finally, by increasing the core density parameter, ρ_c, the value of the spreading length scales are decreasing.

References

1. Frostig, Y., Baruch, M.: Localized load effects in high-order bending of sandwich panels with flexible core. J. Eng. Mech. **122**(11), 1069–1076 (1996)
2. Shen, H.-S., Li, S.-R.: Postbuckling of sandwich plates with FGM face sheets and temperature-dependent properties. Composites: Part B **39**, 332–344 (2008)

3. Zhao, J., Li, Y., Ai, X.: Analysis of transient thermal stress in sandwich plate with functionally graded coatings. Thin Solid Films **516**, 7581–7587 (2008)
4. Allen, H.G.: Analysis and Design of Structural Sandwich Panels. Pergamon Press, London (1969)
5. Plantema, F.J.: Sandwich Construction. Wiley, New York (1966)
6. Petras, A., Sutcliffe, M.P.F.: Indentation resistance of sandwich beams. J. Compos. Struct. **46**, 413–424 (1999)
7. Frostig, Y., Baruch, M., Vilnay, O., Sheinman, I.: High-order theory for sandwich-beam behaviour with transversely flexible core. J. Eng. Mech. (ASCE) **118**(5), 1026–1043 (1992)
8. Frostig, Y., Baruch, M.: Localized load effects in high-order bending of sandwich panels with flexible core. J. Eng. Mech. **122**(11), 1069–1076 (1996)
9. Khalili, M.R., Malekzadeh, K., Mittal, R.K.: A new approach in static and dynamic analysis of composite plates with different boundary conditions. J. Compos. Struct. **169**, 149–155 (2005)
10. Malekzadeh, K., Khalili, M.R., Mittal, R.K.: Local and global damped vibrations of sandwich plates with a viscoelastic soft flexible core: an improved high-order approach. J. Sandw. Struct. Mater. **7**(5), 431–456 (2005)
11. Chi, S.H., Chung, Y.L.: Mechanical behavior of functionally graded material plates under transverse load—Part I: Analysis. Int. J. Solids. Struct. **43**, 3657–3674 (2006)
12. Mindlin, R.M.: Influence of rotary inertia and shear on flexural motions of isotropic elastic plates. J. Appl. Mech. **18**, 31–38 (1951)
13. Malekzadeh, K., Khalili, M.R., Olsson, R., Jafari, A.: Higher-order dynamic response of composite sandwich panels with flexible core under simultaneous low-velocity impacts of multiple small masses. J. Solids Struct. **43**, 6667–6687 (2006)

Micromechanical Fibre-Recruitment Model of Liquid Crystalline Polymer Reinforcing Polycarbonate Composites

K. L. Goh and L. P. Tan

Abstract Injection-molded in situ anisotropic liquid crystalline polymeric (LCP) micro-fibres reinforcing polycarbonate (PC) composite are light-weight materials with 'tailorable' mechanical properties. Using a novel fibre-recruitment (FR) computer model, we have investigated the effects of compatibilization on the microstructure–property relationship in LCP-PC composites, focusing on the recruitment, pull-out and rupture of LCP fibres and LCP-PC interfacial failure. The model represents a parallel array of LCP fibres, of differing lengths and diameters to account for the natural variation, embedded in PC matrix. When an increasing external load acts on the composite, the fibres are recruited in tension. Initially, these fibres undergo linear elastic deformation. At a certain higher applied load, a fraction of the fibres yields and undergoes plastic deformation; eventually, a fraction of these fibres fractures. The FR model was used to evaluate the macroscopic structural and material properties of the fibre, such as fibre diameter, elastic modulus, yield and rupture (σ_r) stresses.

List of Symbols

A_f Cross-sectional area of a LCP fibre
Δ Standard deviation of the mean value
D_f Diameter of the LCP fibre
$\langle D_f \rangle$ Mean diameter of a sub-population of the LCP fibre
Δ Extension of the LCP fibre

K. L. Goh (✉)
School of Engineering, Monash University, Jalan Lagoon Selatan,
Bandar Sunway, 46150 Selangor, Malaysia
e-mail: kheng-lim.goh@newcastle.ac.uk

L. P. Tan
School of Materials Science & Engineering,
Nanyang Technological University, Singapore, Singapore

ε_m Strain in the polycarbonate (PC) matrix at the point if the LCP fibre were absence
ε_f Strain in the LCP fibre
E_f Elastic modulus of the LCP fibre
F Force acting on the LCP fibre
K Boltzmann's constant
k_f Linear stiffness of the LCP fibre
$\langle k_P \rangle$ Mean linear stiffness of the LCP fibre
L Half-length of LCP fibre
LCP Liquid crystal polymer
n The number of LCP fibres used in the simulation
PC Polycarbonate
P_f Plastic modulus of the LCP fibre
q LCP fibre aspect ratio, which is defined as the ratio of the length ($2L$) to the diameter ($D_f = 2r_f$) of the fibre
R Distance from the centre of a LCP fibre to half-way between the gap of the neighbouring fibre
r_f Radius of the LCP fibre
σ_f Fibre axial stress as a function of Z
$\langle \sigma_f \rangle$ Average axial stress in the LCP fibre
$\sigma_{f,\max}$ Maximum axial stress in the LCP fibre
σ_r Fracture stress of LCP fibre
σ_y Yield stress of LCP fibre
T Simulated annealing temperature
τ Interfacial shear stress as a function of Z
χ^2 The goodness-of-fit parameter for the simulated annealing algorithm
Z Normalised distance ($= z/L$) along the fibre axis from the fibre centre to the fibre end, where z is the axial coordinate of the cylindrical coordinate system

1 Introduction

For the past two decades, there has been a growing interest in thermoplastic composites reinforce by liquid crystalline polymer (LCP) fibres [1, 5, 12–14, 16, 18, 23–27]. LCP fibres of high aspect ratio (slenderness) may be self-assembled from p-hydroxybenzoic acid (PHB; Fig. 1b) and poly(ethylene terephthalate) (PET; Fig. 1c). PHB is a phenolic derivative of benzoic acid; it is a white crystalline solid with a high thermal stability; PET is a thermoplastic polymer resin of the polyester family [3]. In particular, Tan et al. [19–22] have carried out many studies on thermoplastic polycarbonates (PC) reinforces with in situ fibres of LCP, comprising 80/20% of PHB and PET. Polycarbonates (PC) possess functional

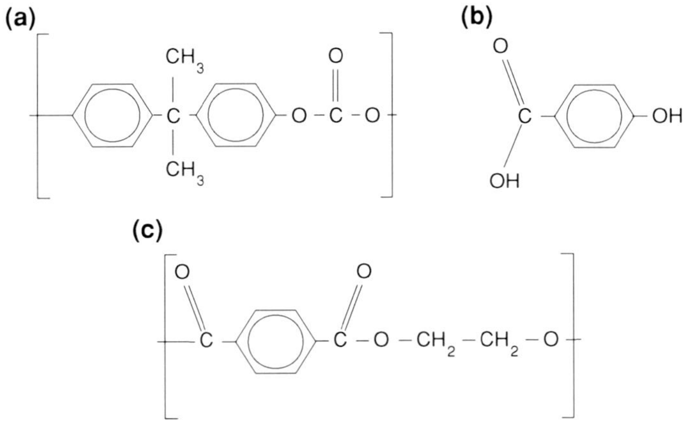

Fig. 1 Chemical structures of (**a**) polycarbonate (PC; repeating unit), (**b**) p-hydrobenzoic acid (PHB) and (**c**) poly(ethelene terephthalate) (PET; repeating unit)

groups linked together by carbonate groups (-O-(C=O)-O-) (Fig. 1a) in a long molecular chain [3]. PC possesses excellent mechanical properties and thermal stability; while PET has similar tensile strength to PC it is also more flexibile with a lower thermal stability than PC [3]. Blending PC with PET compound extends the economic usefulness of PC for making products with a wider range of properties, including those that can withstand high impact or heat [3]. These products range from helmets, power tool housings, battery cases to bottles. Injection-molded LCP-PC blends is one of the most widely used methods; thus it will be useful to find new effective ways to develop LCP blends with 'tailorable' mechanical properties by injection molding [19].

A key issue concerning the synthesis of LCP reinforcing thermoplastic composite is compatibilization [6, 7, 17]. Tan et al. [19–22] implemented compatibilization by trans-esterification of PC and LCP with the aid of a catalyst. It follows that long slender LCP fibres can be crystallized by optimizing the following key operating parameters: shear (flow) rate for extrusion, viscosity ratio and LCP content. In particular, high shear rate allows for the production of long and slender fibres which, upon crystallization, yields LCP-PC composites with high strength and toughness [20]. On the other hand, shear flow increases the temperature of the melt by inducing localized sites of high temperature around the fibres. Consequently the fibres relax into globular structures [21, 22]. One approach to minimize relaxation would be to cool the blend rapidly [21, 22]. In this case, the compatibilizer provides two key advantages: it enhances the mechanical interactions between the LCP fibres and PC matrix and it minimizes fibre relaxation by elevating the critical temperature for this to occur [21, 22].

This chapter describes an investigation on the effects of compatibilizer on the microstructure–property relationship in injection-molded LCP-PC composite using a novel fibre-recruitment (FR) model. A key motivation for this study is that the

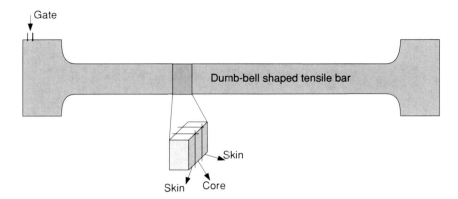

Fig. 2 Schematic of the cross-section of a LCP-PC composite in a dumb-bell shaped mold (ASTM M-II standard). Here, two key regions, known as 'skin' and 'core', are indicated. The dumb-belled shaped tensile bar was produced by injecting palletized LCP-PC dried extrudate, via the gate, into the mold

predictions from the FR model could be useful for optimizing the synthesis parameters, namely shear rate and temperature of melt. The FR model was adapted for studying compatilized and uncompaibilized composites according to the shear-lag and shear-sliding theories, respectively, for discontinuous fibres. The model was evaluated by fitting to experimental stress–strain data, derived from tensile testing of compatibilized and uncompatibilized LCP-PC composites in earlier studies [19–22], to derive the fibre structural and mechanical properties. For preparing the compatibilizer lanthanum acetylacetonate hydrate was used in the catalytic trans-esterification process, aided by an inhibitor [19–22]. Figure 2 shows the synthesis of a dumb-bell shaped LCP-PC tensile bar by injection-molding [21].

2 Micromechanical Fibre-Recruitment Model

2.1 Shear–Lag Model of Compatibilized Composites

Consider a parallel array of discontinuous LCP fibres embedded in and bonded to PC matrix in a compatibilized composite. Here, the fibre axis defines the z axis of the cylindrical coordinate system and the fibre centre defines the origin of the coordinate system so that $Z (= z/L)$ represents a normalized coordinate with values ranging from 0 to 1. We approximate the LCP fibres as uniform cylinders; here r_f and $2L$ represent the fibre radius and length, respectively. The cross-sectional area of a fibre is given by $A_f = \pi r_f^2$. Let R represents the centre-to-center distance between this fibre and adjacent fibre. When the composite is subjected to an external applied load acting in the direction of the fibre axis, more and more fibres are recruited into tension from their relaxed state. Since compatibilization increases the interfacial adhesion and miscibility between the LCP and PC

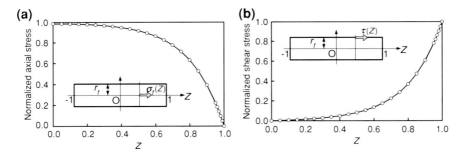

Fig. 3 Graphs of normalized (**a**) axial stress (σ_f/σ_{max}) and (**b**) interfacial shear stress (τ/σ_{max}) versus Z along the LCP fibre from the centre ($Z = 0$) to the tip ($Z = 1$) of the fibre. Inset in each graph is a schematic of the LCP fibre

components [19], stress transfer between LCP fibres bonded to PC matrix in the composite may be described by a shear–lag model [4]. Here, the PC matrix deforms elastically and shear–lags over the LCP fibres; as each LCP fibre deforms elastically, stress is generated within the fibre. Thus, stress is transferred from the matrix to the fibres via the bonds at the fibre-matrix interface. Consider the axial strains in a fibre; let ε_m represents the corresponding strain the PC matrix would undergo (at the same point if the fibre were absent). Also let G_m and E_f represent the shear modulus of the matrix and elastic modulus of the fibre, respectively. The axial stress generated at any point Z, i.e. within the fibre, is given by

$$\sigma_f(Z) = \sigma_{max}\left[1 - \frac{\cosh(\beta Z)}{\cosh(\beta L)}\right] \quad (1)$$

where

$$\beta = \sqrt{\frac{2\{G_m/E_f\}}{r_f^2 \ln(R/r_f)}} \quad (2)$$

and

$$\sigma_{max} = \varepsilon_m E_f. \quad (3)$$

A typical plot of σ_f/σ_{max} versus Z is shown in Fig. 3a. Thus, we find that $\sigma_f \sim \sigma_{max}$ at $Z = 0$. The axial stress remains unchanged over a large proportion of the central region of the fibre; towards the fibre tip, the stress decreases non-linearly and rapidly until it becomes zero at $Z = 1$.

In Eq. 1, the subscript in σ_f describes the stress generated in the fiber. Thus, we find that the macroscopic stress, i.e. the average value of σ_f, i.e. $\langle\sigma_f\rangle$, is related to the local strain state in the matrix, i.e. ε_m, given by

$$\langle\sigma_f\rangle = \varepsilon_m E_f \left[1 - \frac{\tanh(\beta L)}{\beta L}\right]. \quad (4)$$

We note that the right-hand side of Eq. 4 is weighted by $E_f[1-\tanh(\beta L)/(\beta L)]$. The shear stress ($\tau$) acting tangentially at the fibre-matrix interface also varies with Z. From Fig. 3b, we find that τ is zero at the fibre centre (i.e. $Z = 0$) but increases non-linearly, rapidly to a maximum value at the fibre end (i.e. $Z = 1$). In this case, the maximum shear stress (τ_{max}) is given by

$$\tau_{max} = \sigma_{max}\sqrt{\frac{G_m/E_f}{2\ln(R/r_f)}}\tanh(\beta L). \qquad (5)$$

To quantify the stress uptake by the LCP fibres, we define a dimensionless quantity known as the stress transfer ratio, σ_f/τ. Note that in Kelly and MacMillan [11], the stress transfer ratio was defined as τ/σ_f. Nevertheless, both quantities are used to model the same concept; we have used the inverse of what was defined by Kelly and MacMillan [11] because τ/σ_f yields a very small value and so it is not straight-forward to interpret. Letting $\sigma_f/\tau = \sigma_{max}/\tau_{max}$, from Eq. 5 we have

$$\sigma_f/\tau = \sqrt{\frac{2\ln(R/r_f)}{G_m/E_f}}\coth(\beta L). \qquad (6)$$

Thus, the stress transfer depends on the moduli of the fibre and matrix, i.e. E_f and G_m, respectively, the fibre size (i.e. r_f) and fibre–fibre interaction distance (i.e. R).

We note that a varying τ as a function of Z may be justified at the molecular level as follows. There is no interaction between the compatibilized PC matrix macromolecules and the LCP fibre surface around the fibre surface at $Z = 0$ where the mirror symmetry axis of the fibre is defined. However, the number of interactions (presumably dominated by covalent bonds) per unit area of the interface is expected to increase as we consider distances away from the centre towards the fibre end. Shear of the interface involves overcoming these intermolecular covalent forces at the interface. An increasing area density of interactions may be possible if the macromolecular composition of the PC matrix and the LCP fibre varies along the length of the fibre.

2.2 Shear-Sliding Model of Uncompatibilized Composites

Similar to the previous case, we consider a parallel array of discontinuous LCP fibres embedded in PC matrix but in an uncompatibilized composite. In this case, we may assume that there is no adhesion between the LCP and PC components; the LCP fibres are held in PC matrix by residue stresses. As the load acting on the composite increases, the PC matrix deforms plastically and shear-slides over the LCP fibres as they deform elastically. Stress is transferred from the matrix to the fibres via a constant frictional shear stress (τ) acting tangentially at the fibre–matrix interface. According to Kelly and MacMillan [11], the stress transfer

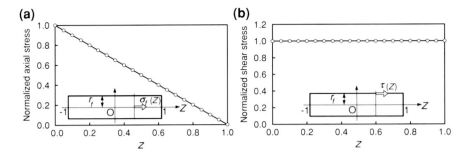

Fig. 4 Graphs of normalized (**a**) axial stress ($\sigma_f/q\tau$) and (**b**) interfacial shear stress ($\tau/q\sigma_{max}$) versus Z along the LCP fibre from the centre ($Z = 0$) to the tip ($Z = 1$) of the fibre

mechanism in uncompatibilized LCP-PC composites is described by a shear-sliding model. The local stress state in each fibre is given by

$$\sigma_f(Z) = 2\tau q[1 - Z], \quad (7)$$

where q denotes the fibre aspect ratio, i.e.

$$q = L/r_f, \quad (8)$$

L, r_f, and Z are parameters which have been defined in the previous section.

A plot of normalized axial stress ($\sigma_f/\tau q$) versus Z is shown in Fig. 4a. The maximum value of σ_f, i.e. $\sigma_{f,max}$ ($= 2\tau q$) occurs at $Z = 0$; thereafter, the stress decreases linearly until it becomes zero at the fibre end ($Z = 1$). Figure 4b shows graphical representation of the distribution of τ along the fibre surface with respect to Z; the constant τ argument addresses, at the molecular level, a constant number of interactions per unit area between the PC matrix macromolecules and the LCP fibre surface throughout the fibre length. Shear of the interface involves overcoming these intermolecular forces at the interface. A constant number of interactions per unit area may be possible if the macromolecular composition of the PC matrix and the LCP fibre do not vary along the length of the fibre.

From Eq. 8, we find that the stress transfer ratio for the shear-sliding model is described by

$$\sigma_f/\tau = 2q. \quad (9)$$

Here, Eq. 9 characterizes the stress transfer between the interfacial shear stress and the axial stress, modulated by q, for the shear-sliding model. Note that the macroscopic stress, given by

$$\langle \sigma_f \rangle = \tau q, \quad (10)$$

is related to the local fibre-matrix interfacial stress.

2.3 Stress–Strain Relationship

This section explains the fibre stress–strain relationship in compatibilized and uncompatibilized LCP-PC composites. We can model the macroscopic force generated in the fibre, F_f, in response to the external load, by relating F_f to the macroscopic stress, given by

$$F_f = \langle \sigma_f \rangle A_f. \tag{11}$$

When the fibre responses elastically to the external load, according to Hooke's law for elasticity, the relationship between F_f and the macroscopic fibre deformation (Δ_f) may be modeled using

$$F_f = k_f \Delta_f, \tag{12}$$

where k_f is the elastic spring constant. We can also express Eq. 12 in terms of $\langle \sigma_f \rangle$ ($= F_f/A_f$) and macroscopic fibre strain, ε_f ($= \Delta_f/L$), as

$$\langle \sigma_f \rangle A_f = k_f \varepsilon_f L_f, \tag{13}$$

or

$$\langle \sigma_f \rangle = E_f \varepsilon_f, \tag{14}$$

where the elastic modulus, E_f, is given by

$$E_f = k_f L_f / A_f. \tag{15}$$

Equation 15 expresses a linear stress–strain relationship that applies to the fibre as it responses elastically to the applied load. Since the magnitude of E_f depends on the nature of the fibre, it is expected that the stiffness of fibres within compatibilized composites will be different from uncompatibilized composites because the process of compatibilization is also responsible for altering the nature of the LCP fibres.

2.4 Failure Mechanisms

This section describes the failure mechanisms in the compatibilized and uncompatibilized LCP-PC composites. Here the fibre yields as the load acting on the fibre increases up to a point such that $\langle \sigma_f \rangle$ is greater than the macroscopic stress at yielding (σ_y). It follows that the relationship between the macroscopic stress and strain is modeled by a linear-plastic equation,

$$\langle \sigma_f \rangle = P_f \varepsilon_f, \tag{16}$$

where P_f is the fibre modulus during yielding, which is quantified by the slope of the stress–strain curve after yield. Eventually, the fibre fractures completely as the load increases up to a point such that $\langle\sigma_f\rangle$ is greater than the macroscopic rupture stress of the fibre (σ_r). Owing to compatibilization, which alters the nature of the LCP fibres and PC matrix, the magnitudes of P_f, σ_y and σ_r from compatibilized composites may differ from uncompatibilized composites.

3 Computer Simulation

3.1 Model Description

The shear–lag and shear-sliding models were implemented by a computer algorithm (MatLab, version 7) which evaluates q, E_f, P_f, σ_y and σ_r by fitting, using a simulated annealing approach (see the next section), to experimental data derived from earlier reports on compatibilized and uncompatibilized LCP-PC composite tensile bars [19]. A flow-chart of the algorithm, hereafter known as the fibre recruitment (FR) algorithm, is shown in Fig. A.1 (Appendix).

The parallel array of LCP fibres features varying diameters; this is to account for the natural variation. It follows that the overall fibre diameter distribution comprises two normally distributed sub-populations with means, $\langle D_f \rangle_1$ and $\langle D_f \rangle_2$ (where $\langle D_f \rangle_1 < \langle D_f \rangle_2$) and standard deviations (SD), $\delta_{D,1}$ and $\delta_{D,2}$ (see next section for further details on how this conclusion was arrived). Bar-charts of normalized frequency versus diameters are shown in Fig. 5; an optimization procedure, using the simulated annealing approach (see next section), was used to generate these distributions. Here, we noted that the most frequent number of fibres in the compatibilized and uncompatibilized sample are $D_f \sim 3.3$ and 14 μm, respectively; for further details, see Tan et al. [19, 20]. The natural variation also extends to length; in this case, q (Eq. 8 will vary from one fibre to another. Thus, q is assumed to follow a normal distribution with mean, $\langle q \rangle$, and SD, δ_q; the q values were determined by simulated annealing (i.e. next section).

The model of Frisen et al. [8] was adapted for the purpose of simulation by considering the simultaneous mechanical response of the fibres to an external load (Fig. 6). When an increasing external load acts on the composite, the PC matrix deforms and all fibres were recruited in tension. Initially, these fibres undergo linear elastic deformation, govern by Eq. 14. At higher load, a proportion of the fibres yield and undergo plastic deformation govern by Eq. 16; a proportion of these rupture eventually when the stress generated in these fibres exceeded σ_r.

For LCP fibres undergoing elastic deformation, followed by plastic deformation, the natural variation in the fibre stiffnesses, E_f, are derived from the corresponding linear stiffness k_E which is assumed to be distributed normally parameterized by mean, $\langle k_E \rangle$, and SD, δ_E, using Eqs. 12–15. The same approach was also implemented to derive P_f, i.e. from the corresponding linear stiffness, k_P (normal distribution parameterized by mean, $\langle k_P \rangle$, and SD, δ_P). Consequently, each fibre takes up stress (σ_f) and strains (ε_f) differently, depending on the stiffness of the fibre.

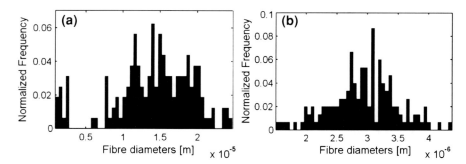

Fig. 5 Fibre diameter frequency distributions of (**a**) uncompatibilised and (**b**) compatibilised LCP-PC composites

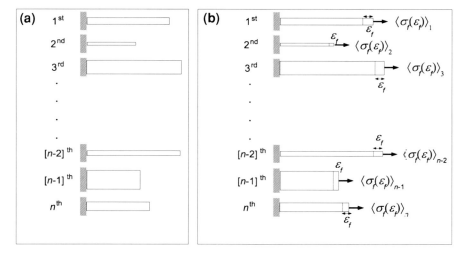

Fig. 6 Fibre-recruitment (FR) model of LCP-PC composite undergoing tensile test, showing linear elements (representing the fibres) in action: (**a**) the relaxed fibres before the application of an external load; (**b**) during loading. Here, the fibres are numbered from 1 to n, where n is the total number of fibres in the PC matrix, $\langle \sigma_f \rangle_i$ is the macroscopice stress generated in the ith fibre (where i = 1, 2, 3, ...,n) in response to an external load and ε_f is the fibre strain

3.2 Simulated Annealing

A simulated annealing algorithm was incorporated into the FR model (Fig. 1) to determine q, E_f, P_f, σ_y and σ_r by linear least-square fitting to experimental stress–strain data of compatibilized and uncompatibilized LCP-PC composite tensile bars derived from an earlier report [19]. The method of simulated annealing was chosen because it is a robust technique for optimization problems of large scale problems, especially ones where a desired global extremum is hidden among many, poorer, local extrema; this method has been applied successfully to a wide range of problems, such as the determination of fibre diameters from small angle X-ray scattering data [10] to name one.

To implement the algorithm we defined an objective function to be minimized given by the (chi square) goodness-of-fit parameter (χ^2) for quantifying the optimal (linear least squares) fit of the model to the experimental data. The space over which this function is defined is a discrete, but very large, configuration space containing the set of possible normal frequency distributions of D_f, q, E_f, P_f, σ_y and σ_r. A Matlab procedure, incorporating the random number generator, was used to generate these distributions; the number of bins was set equal to the number of fibres used in the simulation model for generating these distributions. A simulated thermodynamic system was assumed to change its configuration from χ_1^2 to χ_2^2 with probability $P = \exp(-[\chi_2^2 - \chi_1^2]/kT)$, where k is Boltzmann's constant and T is the simulated annealing temperature. If $\chi_2^2 < \chi_1^2$, the probability is greater than unity; in such cases the change is arbitrarily assigned a probability $P = 1$, i.e., the system always took such an option. This general scheme, of always taking a downhill step while sometimes taking an uphill step, is known as the Metropolis algorithm. Thus, implementing the Metropolis algorithm for the FR model required the following input parameters,

1. The possible system configurations, i.e. the normal distributions of D_f, q, E_f, P_f, σ_y and σ_r (note that both D_f and q were constrained by the range of values observed from experiments [19, 20]);
2. A generator of random changes in the configuration;
3. An objective function χ^2 whose minimization is the goal of the procedure;
4. A control parameter T, obtained by trial-and-error, and an annealing schedule which tells how it is lowered from high to low values, e.g., after how many random changes in configuration is each downward step in T taken, and how large is that step. In the trial and error approach, initially we generated some random rearrangements, and use them to determine the range of values of trial-and-error that will be encountered from move to move. Choosing a starting value for T which was considerably larger than the largest $\chi_2^2 - \chi_1^2$ normally encountered, we proceeded downward in multiplicative steps each amounting to a 10 percent decrease in T. We held each new value of T constant for $100 N$ reconfigurations, or for $10 N$ successful reconfigurations, whichever came first (N represents the number of fibres). When efforts to reduce χ^2 further become sufficiently discouraging, we stopped the simulation for T.

In addition to our starting set of configurations of q, E_f, P_f, σ_y and σ_r, the other input parameters that were required for the simulated annealing model include: the number of fibres and the nature of the fibre diameter distribution. Here, $N = 150$ fibres for the uncompatibilized and compatibilized samples; these are the mimimum number that is shown to yield an optimal solution for a good fit between the computational and experimental results (corresponding to a composition of PC and LCP of 15 and 85% by weight, respectively). The value of N assigned to both uncompatibilized and compatibilized samples was the same because this study was not intended to model the effects of LCP concentration (which would have an influence on the mechanical properties [20]). A bimodal distribution (arising from two normally distributed subpopulations) was assumed for the underlying nature of the fibre diameter; we arrived at this assumption by evaluating the smallest number of sub-populations that can be used to fit the stress–strain curve. In this case, the ratio of the number of fibres in the distribution with the smaller mean to that of the larger mean was found to be 6:74 and 4:71 for the uncompatibilized and compatibilized samples, respectively. The starting point of our evaluation for the appropriate fibre diameter distributions was based on experimental observations reported in an earlier study [20].

4 Predictions

4.1 Stress–Strain Relationship

The predicted stress–strain curve is plotted together with the experimentally derived curve in Fig. 7 For the uncompatibilized sample, the FR model predicts the yielding of the first and the last fibre correspond to strain values of 0.01 and 0.30, respectively. Thus, the last fibre to yield occurred at the composite fracture strain. The first fibre to rupture occurred at a strain value of 0.04, which lies between the strain values corresponding to the yielding of the first and the last fibre. Also, 61.9% of the fibres yielded; it was found that $\sigma_y = 2.0$ MPa. Not all the fibres ruptures after yielding. We found that 92.9% of the yielded fibres eventually ruptured; in other words, this is 57.5% of the total number of fibres. Rupture of these fibres was found to occur at $\sigma_r = 20.0$ MPa.

However, for the compatibilized sample, the FR model predicts that the yielding of the first and the last fibre correspond to strain values of 0.08 and 0.60, respectively; again, the latter occurred at the composite fracture strain. In addition, the first fibre to rupture occurred at a strain value of 0.36; this lies in between the strain values corresponding to the yielding of the first and the last fibre. The proportion of fibres which yielded is 61.3% which is close to that obtained for the uncompatibilized sample; yielding was found to occur at $\sigma_y = 10.0$ MPa. However, only 69.7% of the yielded fibres eventually ruptured; in other words this is 42.7% of the total fibre. Here, the rupture of these fibres was found to occur at $\sigma_r = 44.0$ MPa.

4.2 Fibre Structure

The distributions of q for the uncompatibilized and compatibilized samples are shown in Fig. 8 The results from these samples reveal a slight skewness to the left, suggesting that the q values are not normally distributed.

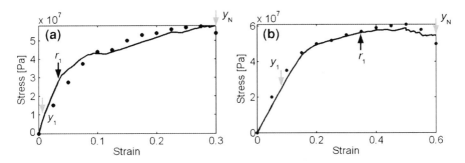

Fig. 7 Stress–strain relationships of (**a**) uncompatibilized and (**b**) compatibilized LCP-PC composites. Here, *arrows* y_1 and y_N are used to indicate the first and last points on the curve where fibre yielding occurs; *arrow* r_1 indicates the point on the curve where the first fibre ruptured

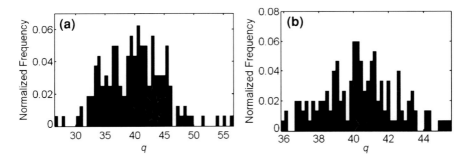

Fig. 8 Frequency distributions of fibre aspect ratio (q) from (**a**) uncompatibilised and (**b**) compatibilised LCP-PC composites

In the case of the uncompatibilized sample, the spread of q values fall within the range 25–56. The most frequent number of fibres corresponds to $q \sim 41$.

In the case of the compatibilized sample, the spread of q values ranges from 36 to 45. Here, the most frequent number of fibres in the compatibilized sample corresponds to $q \sim 40$ which is not very different from the uncompatibilized sample. Note that high temperature was used in the blending process [19]. If the temperature was increased at a constant rate of 2°C/min from 270 to 300°C, Tan et al. [21] found that the aspect ratio and length of the fibres in uncomptabilized composites decreases steadily; at 295°C, the fibres were completely relaxed. In contrast, no appreciable relaxation was observed in the fibres of the compatibilized composites with increasing temperature. Thus, it is less energetically favourable for the fibres in the compatibilized sample to relax [21]; this can be exploited to modulate the q values by narrowing range of slenderness to encompassing a desirable value (Fig. 8b).

The yielded fibres in the uncompatibilized and compatibilized samples are identified by the distributions of fiber diameter shown in Fig. 9. The distributions

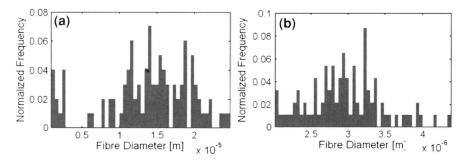

Fig. 9 Frequency distributions of fibre diameter of yielded fibres in the (**a**) uncompatibilised and (**b**) compatibilised LCP-PC composites

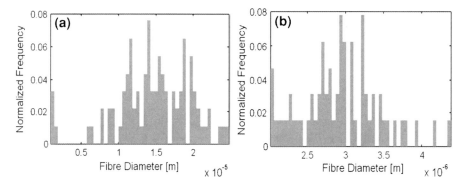

Fig. 10 Frequency distributions of fibre diameter of fractured fibres from (**a**) uncompatibilised and (**b**) compatibilised LCP-PC composites

of the uncompatibilized and compatibilized samples, i.e. the spread of fibre diameters and the profile of the distribution, are not quite the same as those of Fig. 5a and b, respectively. In particular, for the uncompatibilized sample the most frequent number of fibres for yielding corresponds to $D_f \sim 14$ μm. For the compatibilized sample the most frequent number of fibres corresponds to $D_f \sim 3.3$ μm. Within the range of fibre diameters shown in Fig. 9, it is observed that the uncompatibilized sample results in a higher proportion of larger fibre yielding; these are expected to correspond to the ones with globular structures because these structures possess low aspect ratio and hence are not effective in taking up stress from the PC matrix [9]. On the other hand, the compatibilized sample results in a higher proportion of smaller fibre yielding; this is expected to correspond to the ones with high slenderness because these structures are effective in taking up stress from the PC matrix [9].

The distributions of fiber diameter associated with fractured fibres in the uncompatibilized and compatibilized samples are shown in Fig. 10a and b,

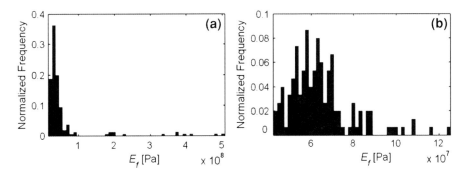

Fig. 11 Frequency distributions of fibre elastic modulus (E_f) from: (**a**) uncompatibilised and (**b**) compatibilised LCP-PC composites

respectively. Note also that the distributions of the uncompatibilized and compatibilized samples, i.e. the spread of fibre diameters and the profile of the distribution, are not quite the same as those of Fig. 9. Again, it is noteworthy to emphasize that within the range of fibre diameters shown in the Fig. 10., in the uncompatibilized sample, fracture occurs in a greater number of fibres with bigger diameter (corresponding to globular structures). In the compatibilized sample, fracture occurs in a greater number of fibres with the smaller diameters (corresponding to high slenderness).

4.3 Fibre Modulus

The uncompatibilized sample exhibits a wide spread of E_f values ranging up to 5×10^8 Pa. The most frequent number occurs in the lower range up to 1×10^8 Pa; the frequency value peaks at slightly less than $E_f \sim 0.5 \times 10^8$ Pa. The compatibilized sample exhibits a narrower spread of E_f values, i.e. ranging up to slightly more than 12×10^7 Pa. The most frequent number occurs in the low to mid range from 4×10^7 Pa to slightly more than 8×10^7 Pa; the frequency value peaks at $E_f \sim 6.0 \times 10^7$ Pa.

Figure 12 shows that the values of P_f are greater than zero, indicating that both compatibilized and uncompatibilized LCP-PC composites exhibit work hardening. The uncompatibilized sample yields a wider spread of P_f values than the compatibilized sample. For the uncompatbilized sample, high frequency values are confined to the lower range of P_f values up to 0.5×10^8 Pa; the highest frequency value peaks at slightly less than 0.25×10^8 Pa. For the compatibilized sample, high frequency values confined to within 0.6×10^7 to 1.0×10^7 Pa; the highest frequency value peaks at slightly less than 1.0×10^7 Pa.

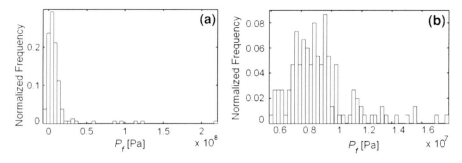

Fig. 12 Frequency distributions of fibre plastic modulus (E_f) from (**a**) uncompatibilised and (**b**) compatibilised LCP-PC composites

5 Discussion

5.1 LCP-PC Interface

Achieving effective interfacial adhesion between the LCP and PC is key to enhancing the mechanical properties. The LCP fibres that were considered in this study may be regarded as two-phase structures of copolyesters of PHB with PET [14]. In this respect, the PHB-PET copolyesters exhibit the following amorphous phases: a PET-rich "flexible" phase and a PHB-rich "rigid" phase. Going by this argument, it is speculated that the PET-rich phase, not the PHB-rich phase, is miscible in PC (although further tests, using differential scanning calorimetry, would be necessary to confirm this). This miscibility ensures that the LCP crystallites, which act as a cross-link within the PC matrix, can provide a more uniform reinforcement throughout the PC matrix. However, as these cross-linking are thermally labile [3], we find relaxation of the LCP fibres occurring within a high temperature environment that is close to its melting point [21]. Introducing compatibilizer to the LCP-PC blend is intended for enhancing interfacial adhesion; it also provides a way for tailoring the mechanical properties. In the studies of Tan et al. [19–22], it is noteworthy to point out that the compatibilizer were block or graft copolymers possessing segments with chemical structures or solubility parameters that are similar to those of the polymers being blended. Acting as polymeric surfactants, the compatibilizer is able to reduce the interfacial tension, which in turn promotes interfacial adhesion, leading to the formation of even more uniform distribution of the dispersed LCP fibres [6]. Consequently an extensive dispersion of LCP fibres and adhesion between LCP and PC lends to improvement in the mechanical properties of the compatibilized blend [6, 7, 17].

5.2 Failure of LCP Fibres

The FR model predicts that the yield and fracture stresses of LCP fibres are higher in the compabilized sample as compared to the uncompatibilized sample. This suggests that the average length of the polymer chains in the LCP fibres of the uncompatibilized sample are shorter than those in the compatibilized since the shorter ones in the former would be relatively more mobile, yielding a reduced contribution to the fibre strength. Molecular alignment is an important factor in LCP fibre formation [3]; improving molecular alignment is aided by the presence of long polymer chains within LCP fibres which yields stronger dipole interaction forces between side groups (when present) of the LCP fibres than hydrogen bonds or van Der Waals forces [3].

5.3 Stiffening Mechanism

Predictions from the FR model show that the elastic modulus of the LCP fibres from the compatibilized sample is smaller than those from the uncompatibilized sample. This arises because there is a decrease in the crystallinity of the compatibilized fibre [3]. It may be said that the compatibilizer is acting like a plasticizer. The plasticizing effect is attributed to the relatively smaller molecular size of the compatibilizer (in comparison to the long polymering chains) located within the LCP fibres; the smaller molecules from the compatibilizer are able diffuse into the outer-most regions of the fibres more effectively than larger oness. Accumulation of these molecules between the long polymeric chains leads to a net repulsive force which draws the long polymeric chains further apart [3] and this lowers the LCP fibre modulus. By the rule of mixture for modulus, it follows that the modulus of the compatibilized LCP-PC composite (Fig. 7a) is smaller (hence more flexible) than the uncompatibilized sample (Fig. 7b).

5.4 Structural Distribution of LCP Fibre

The FR model predicts that the underlying stress–strain profile of the compatibilized and uncompatibilized samples is attributed to the bimodal fibre diameter distribution. It is well-known in connective tissue study [15] that fibre diameter is responsible for the toughening of the tissue. By an analogy to the argument set up by Parry et al. [15], an explanation is provided here to address how fibre diameter influences the strain energy density (equals to the area under the stress–strain curve) and hence the profile of the stress–strain relationship. We speculate that the early part of the stress–strain curve, from the toe up to the end of the linear region, is attributed mainly to the recruitment of stubby fibres (of large diameter and short length), especially those in the core regions. In the latter stages of the loading

Table 1 Upper and lower limits of stress transfer ratio (σ_f/τ) for uncompatibilized and compatibilized composites

	Lower limit of σ_f/τ	Upper limit of σ_f/τ
Uncompatibilized	50	112
Compatibilized	769	192

process, an increasing number of fibres with high slenderness (corresponding to fibres small diameter with long axial length, especially in the skin regions) contribute to the latter half of the linear part of the stress–strain curve. We also note that the strain energy density of the compatibilized sample is larger than that of the uncompabilized sample (Fig. 7). Thus, the larger toughness of the compatibilized sample may be attributed to a higher number of fibres with high slenderness so that more energy may be absorbed for fracturing these LCP fibres.

5.5 Stress Transfer Efficiency

Finally, this paragraph is intended to discuss the effectiveness of compatibilization for stress transfer in the LCP-PC composite. In earlier sections, we have defined the stress transfer ratio as a ratio, i.e. σ_f/τ. Here, we will use the stress transfer ratio of Eqs. 6 and 9 to predict the stress transfer efficiency in uncompatibilized and compatibilized samples. To begin, we note that the Poisson's ratios (v_m) of PC fall within [0.37, 0.38] and that of E_m falls [2.0, 2.5 MPa] [3]. According to the elastic constants relationship,

$$G_m = \frac{E_m}{2[1+v_m]}, \qquad (17)$$

we find that the value of G_m falls within [0.72, 0.91 MPa]. Now, an inspection of Fig. 5. A reveals that the fibre diameters $\in$ (1.0, 4.5 μm); the corresponding radii $\in$ (0.5, 2.3 μm). Next, we define an upper and a lower limit for the values of the inter-fibre distance, i.e. R, corresponding to 1.0 and 4.4 μm, respectively. Using these data, the upper and lower limits of σ_f/τ for the compatibilized and uncompatibilized samples may be easily calculated using Eqs. 5 and 9; the results are listed in Table 1. This simple calculation reveals that compatibilization improves the stress transfer from PC matrix to LCP fibres.

6 Conclusions

Using a novel FR computer model, we have investigated the effects of compatibilization on the microstructure-property relationship in LCP-PC composites. Here, the focus is on the recruitment, pull-out and rupture of LCP fibres and LCP-

PC interfacial failure. The model represents a parallel array of LCP fibres, of differing lengths and diameters to account for the natural variation, embedded in PC matrix. When an increasing external load acts on the composite, the fibres are recruited in tension. Initially, these fibres undergo linear elastic deformation. At a certain higher applied load, a fraction of the fibres yields and undergoes plastic deformation; eventually, a fraction of these fibres fractures.

The FR model was used to evaluate the macroscopic structural and material properties of the fibre, namely D_f, q, E_f, P_f, σ_y and σ_r. A simulated annealing approach was used to fit the FR model to experimental stress–strain data obtained from earlier reports on compatibilized (i.e. by catalytic transesterification) and uncompatibilized LCP-PC composite tensile bars.

The FR model predicts that

1. 63% of the fibres in the compatibilized sample undergo yielding but only 50% of the total number of fibres eventually rupture;
2. while the proportion of yielded fibres in the uncompatibilized sample are similar to that of the compatibilized sample, a higher proportion of the fibres eventually rupture (56%) in the uncompatibilized sample;
3. although the fibres in the compatibilized sample possess values of E_f that range from 50 to 100 MPa, those from the uncompatibilized sample feature a wider spread of E_f values (encompassing that of compatibilized specimen);
4. the uncompatibilized sample suffered from more extensive work hardening (i.e. the slope after yield point was non-zero) than the compatibilized sample.

The predictions from the FR model were used to evaluate the mechanism of stress transfer between the LCP fibres and PC matrix. It is revealed that the stress transfer ratio from the matrix to the fibre in the compatibilized sample is higher than that from the uncompatibilized sample.

Thus, the role of compatibilization is multi-functional:

1. it enhances the stability of the LCP fibre against thermal relaxation, yielding fibres with controllable aspect ratios;
2. it increases the interfacial adhesion and miscibility between the LCP and PC components;
3. it modulates the polymer chain length of the fibres, producing fibres with high yield and rupture stresses;
4. it modulates the crystallinity of the fibre, producing fibres that are highly flexible than those of the uncompatibilised blend.

Acknowledgments The authors would like to thank Unitika (Japan) and Bayer for providing the materials for carrying out the experiments on LCP-PC composites for an earlier study. The authors would also like to acknowledge Monash University for funding the computational work.

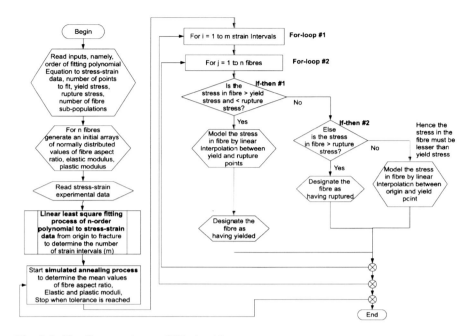

Fig. A.1 The fibre-recruitment (FR) algorithm

Appendix

Figure A.1 illustrates the fibre-recruitment (FR) algorithm that was developed to evaluate the stress transfer mechanism in the polycarbonate (PC) matrix reinforced by liquid crystalline polymeric (LCP) fibres. The FR algorithm comprises two processes, namely the linear least-square fitting process and the simulated annealing process. The former is needed to fit a user-defined n-order polynomial equation to the stress–strain data points starting from the origin and ending at the point of fracture. The equation is used to determine the number of strain intervals (m) between the origin and the point of fracture. The simulated annealing process (Sect. 3.2) oversees to two for-loops, namely for-loop #1 and #2. For-loop #1 directs the 'stretching' simulation by incrementally increasing the strain in the composite from zero until the last strain point. For-loop #2 checks the stress in every fibre using Eqs. 14 and 16 by taking into account the strain in the composite. Thus, within for-loop #2, there are two if–then statements, namely if–then #1 and #2. If if–then #1 is true, it attempts to model the stress in the fibre by a linear interpolation approach between the yield and rupture points. Otherwise, it proceeds to if–then #2. If if–then #2 is true, the fibre is regarded as ruptured. Otherwise, it presumes that the stress in the fibre is within the elastic limit and hence attempts to model the stress in the fibre by a linear interpolation approach between the origin and the yield point.

References

1. Bassett, B.R., Yee, A.F.: A method of forming composite structures using in situ-formed liquid crystal polymer fibers in a thermoplastic matrix. Polym. Compos. **11**, 10–18 (1990)
2. Beery, D., Kenig, S., Siegmann, A.: Structure development during flow of polyblends containing liquid crystalline polymers. Polym. Eng. Sci. **31**, 451–458 (1991)
3. Cowie, J.M.G., Arrghi, V.: Polymers: Chemistry and Physics of Modern Materials. CRC Press, Boca Raton (2008)
4. Cox, H.L.: The elasticity and strength of paper and other fibrous materials. Br. J. Appl. Phys. **3**, 72–79 (1952)
5. Crevecouer, G., Groeninckx, G.: Morphology and mechanical properties of thermoplastic composites containing a thermotropic liquid crystalline polyme. Polym. Eng. Sci. **30**, 532 (1990)
6. Datta, A., Chen, H.H., Baird, D.G.: The effect of compatibilization on blends of polypropylene with a liquid-crystalline polymer. Polymer **34**, 759–766 (1993)
7. Dutta, D., Weiss, R.A., He, J.S.: Compatibilization of blends containing thermotropic liquid crystalline polymers with sulfonate ionomers. Polymer **36**, 429–435 (1996)
8. Frisen, M., Magi, M., Sonnerup, L., Viidik, A.: Rheological analysis of soft collagenous tissue part 1: Theoretical considerations. J. Biomech. **2**, 13–20 (1969)
9. Goh, K.L., Aspden, R.M., Mathias, K.J., Hukins, D.W.L.: Finite-element analysis of the effect of material properties and fibre shape on stresses in an elastic fibre embedded in an elastic matrix in a fibre-composite material. Proc. R. Soc. Lond. A **460**, 2339–2352 (2004)
10. Goh, K.L., Hiller, J., Haston, J.L., Holmes, D.F., Kadler, K.E., Murdoch, A., Meakin, J.R., Wess, T.J.: Analysis of collagen fibril diameter distribution in connective tissues using small-angle X-ray scattering. Biochimica et Biophysica Acta—Gen. Subj. **1722**, 183–188 (2005)
11. Kelly, A., Macmillan, N.H.: Strong Solids, 3rd edn. Oxford University Press, Oxford (1986)
12. Khan, K.A., Kahraman, R., Hamad, E.Z., Ali, S.A., Hamid, S.H.: Studies on a terephthalic acid and dihydroxydiphenyl sulfone liquid crystalline copolymer and its composites with different thermoplastics. J. Appl. Polym. Sci. **64**, 645–652 (1997)
13. Kiss, G.: In situ composites: Blends of isotropic polymers and thermotropic liquid crystalline polymers. Polym. Eng. Sci. **27**, 410–423 (1987)
14. Lin, Q.H., Jho, J.Y., Yee, A.F.: Effect of drawing on structure and properties of a liquid crystalline polymer and polycarbonate in situ composite. Polym. Eng. Sci. **33**, 789–798 (1993)
15. Parry, D.A.D., Barnes, G.R.G., Craig, A.S.: A comparison of the size distribution of collagen fibrils as a function of age and a possible relationship between fibril size distribution and mechanical properties. Proc. R. Soc. Lond. A **203**, 305–321 (1978)
16. Petrovic, Z.S., Farris, R.J.: Morphology and properties of fibers based on polycarbonate/liquid crystalline polymer blends. J. Polym. Adv. Technol. **16**, 91–99 (1995)
17. Stachowski, M.J., DiBenedetto, A.T.: The effect of block structure of an ethylene terephthalate/hydroxybenzoate copolymer on its ability to form compatible blends. Polym. Eng. Sci. **37**, 252–260 (1997)
18. Seo, Y.S., Hong, S.M., Hwang, S.S., Park, T.S., Kim, U.K., Lee, S.M., Lee, J.L.: Compatibilizing effect of a polyesterimide on the properties of the blends of polyetherimide and a thermotropic liquid crystalline polymer- 2- morphology and mechanical properties. Polymer **36**, 525–534 (1995)
19. Tan, L.P., Yue, C.Y., Tam, K.C., Lam, Y.C., Hu, X.: Effect of compatibilization in injection-molded polycarbonate and liquid crystalline polymer blend. J. Appl. Polym. Sci. **84**, 568–575 (2002)
20. Tan, L.P., Yue, C.Y., Tam, K.C., Lam, Y.C., Hu, X.: Effects of shear rate, viscosity ratio and liquid crystalline polymer content on morphological and mechanical properties of polycarbonate and LCP blends. Polym. Int. **51**, 398–405 (2002)

21. Tan, L.P., Joshi, S.C., Yue, C.Y., Lam, Y.C., Hu, X., Tam, K.C.: Effect of shear heating during injection molding on the morphology of PC/LCP blends. Acta Mater. **51**, 6269–6276 (2003)
22. Tan, L.P., Yue, C.Y., Tam, K.C., Lam, Y.C., Hu, X., Nakayama, K.: Relaxation of Liquid-Crystalline Polymer Fibers in Polycarbonate–Liquid-Crystalline Polymer Blend System. J. Polym. Sci. Part B: Polym. Phys. **41**, 2307–2312 (2003)
23. Tjong, S.C., Meng, Y.Z.: The effect of compatibilization of maleated polypropylene on a blend of polyamide-6 and liquid crystalline copolyester. Polym. Int. **42**, 209–217 (1997)
24. Weiss, R.A., Huh, W., Niclois, L.: Novel reinforced polymers based on blends of polystyrene and a thermotropic liquid crystalline polymer. Polym. Eng. Sci. **27**, 684–691 (1987)
25. Wiff, D.R., Weinert, R.J.: Blending thermotropic liquid crystal and thermoplastic polymers for micro-reinforcement. Polymer **39**, 5069–5073 (1998)
26. Xu, Q.W., Man, H.C., Lau, W.S.: Melt flow behavior of liquid crystalline polymer in situ composites. J. Mater. Process. Technol. **63**, 519–523 (1997)
27. Yi, X.S., Zhao, G.M., Shi, F.: Study on the miscibility and rheological properties of thermotropic liquid crystalline polymers in thermoplastic matrices. Polym. Int. **39**, 11–16 (1996)

Moisture Absorption Effects on the Resistance to Interlaminar Fracture of Woven Glass/Epoxy Composite Laminates

X. J. Gong, K. J. Wong and M. N. Tamin

Abstract The influence of moisture absorption on the interlaminar fracture behaviour of 8/8 harness satin weave glass/epoxy composite was investigated. Two series of specimens with 0°/0° and 90°/90° predominant interfaces immersed in water for different duration were tested under double cantilever beam (DCB mode I), single leg bending (SLB mode I + II) and end notched flexural (ENF mode II) loadings. In general, the apparent flexural modulus: E, and the fracture toughness: G_C, decrease with increasing moisture content. This effect is more remarkable if mode II participation is bigger. The value of G_C measured on 90°/90° specimens reveals higher than that on 0°/0° ones, but the variation in G_C is inversed under ENF loading. The experimental results have been correlated with a criterion previously proposed by the first author expressed by: $G_{TC} = G_{IC} + (G_{IIC} - G_{IC})\left(\frac{G_{II}}{G_I + G_{II}}\right)^m$. A good agreement is shown with $m = 2/3$ at all moisture contents and interfaces. Regarding the R-curves under DCB loading, water absorption leads to a higher rate of increment in the resistance in the early crack growth. However, the maximum of the resistance to the crack growth decreases with the moisture content.

Keywords Woven fibre reinforced composite · Moisture effect · Delamination · Fracture toughness · Mixed-mode ratio

---

X. J. Gong (✉) · K. J. Wong
Département de Recherche en Ingénierie des Véhicules
pour l'Environnement, Université de Bourgogne, 58000 Nevers, France
e-mail: xiao-jing.gong_isat@u-bourgogne.fr

K. J. Wong · M. N. Tamin
Faculty of Mechanical Engineering, Centre for Composites,
Universiti Teknologi Malaysia, 81310 UTM Skudai, Johor, Malaysia

1 Introduction

In recent decades, there is a rapid growth in the use of fibre reinforced polymer (FRP) composites in advanced applications such as aerospace, aeronautics, automobile and marine industries. The major advantages of polymeric composites include high specific stiffness and strength. However, the structures in composite laminate have comparatively lower interlaminar strength; delamination is generally recognised as one of the most common and early detected damage mechanism. Moreover, composite structures are usually subjected to different thermal and moisture environments. The environmental effects could alter the properties of the matrix and the fibre–matrix interface [1], and hence the performance of the composites in terms of strength, stiffness and also damage progression, in particular, the reduction of the resistance to interlaminar fracture has been shown significant. Besides, the difference in the thermal and moisture expansion coefficients between the constituents of the composite induce additional internal stresses, which have to be superposed to the mechanical applied stresses, and therefore vary the behaviour of fibre–matrix interface. Consequently, it is of great importance for better understanding the effects of these factors, which have to be included in the design to predict correctly the lifetime of the composite structures.

Based on fracture mechanics approach, the resistance of composites to delamination is usually characterised by the fracture toughness for the crack initiation, in terms of the critical strain energy release rate, G_C, and by the R-curve for the crack propagation. It is shown dependent not only on environment conditions, but also on loading mode.

Zenasni and his co-workers studied the effects of hygrothermal and hygrothermomechanical on the fracture resistance of woven fibres (2/2 twill glass, 8-harness satin glass and 8-harness satin carbon fibres) reinforced polyetherimide (PEI) composites under double cantilever beam (DCB) mode I and end notched flexure (ENF) mode II loadings [2]. Specimens were aged at 0, 30, 60, 120 and 180 days and then tested at constant temperature and given relative humidity. Results showed that with moisture exposure up to 180 days, the G_{IC} values for 2/2 twill glass and 8H satin carbon fibres reduced, but remained nearly constant for 8H satin glass fibre case. Regarding the value of G_{IIC}, at the end of the ageing, reduction was observed for 2/2 twill glass and 8H satin carbon fibres; on the contrary improvement was found for 8H satin glass fibre composite.

In the studies of Huang and Sun [3], the effects of moisture absorption in 2-layer glass fabric reinforced unsaturated polyester composites were investigated at different durations of immersion. The peeling strength of the composites was found increasing with the immersion time up to 14 days. It is believed that the water molecules seeped into the cracks and voids of the laminates so as to enhance the peeling resistance. Another recent research has investigated the delamination behaviour of carbon/epoxy composite reinforced by thermoplastic particulate [4]. Both dry and wet specimens were tested at low, room, elevated and high (for dry specimens only) temperatures. The specimens were tested using DCB, ENF and single leg bending (SLB) tests. Results indicated that temperature and moisture absorption enhanced the ductility of the matrix, where mode I fracture toughness was improved. However, adverse effect was found on mode II fracture resistance. The mixed-mode fracture toughness increased with moisture but decreased with temperature.

In order to improve delamination behaviour of carbon/epoxy laminates under environment conditions, Walker and Hu [5] proposed to introduce Short Fibre Reinforcement (SFR) into the delamination region. It is shown that the use of SFR enhanced the interlaminar bonding under DCB loading. The improvement of fracture toughness was observed not only for dry specimens, but also for hot/wet specimens. In some cases, the fracture toughness of hot/wet specimens was higher than the dry ones.

Furthermore, it has also been reported that delamination behaviour depends on the fibre orientation of adjacent layers to crack plane. It was indicated that adjacent ply orientations of 0°, 15°, 30° and 45° induced different fracture surface morphology, and thus varied the fracture resistance of the composites [6–9]. However, the trend was not conclusive, as the fracture toughness could be increasing [10], decreasing [11] or even invariable [12]. Gong et al. studied the effects of the stacking sequence, dimensions of specimens and adjacent as well as sub-adjacent fibre orientations on the mode I fracture toughness of glass and carbon/epoxy composites [13]. Results revealed significant influence of these parameters on G_{IC}. In another research, the effects of moisture concentration and edge finishing type on tensile fracture strength were investigated by testing the specimens having the same thickness and three different quasi-isotropic (QI) stacking sequences in carbon/epoxy [14]. It was shown that the stacking sequence has less effects on the moisture absorption process. However the reduction in the tensile fracture strength due to moisture absorption was found evident because of interface degradation.

Regarding the crack propagation, the R-curve describes the variation of the resistance to delamination as a function of the stable crack growth. Under DCB mode I loading, a steep increase of strain energy release rate, G_p, is usually observed in the early crack growth from film insert [11, 14, 15]. And then, the value of G_p tends to a stable level as the crack propagates in a local unstable stick–slip manner. Sometimes in multidirectional (MD) specimens, R-curves increase continually with the crack growth for a long time; the crack was jumping from one interface to another, where different morphology of fracture surface was observed [7, 8]. In this case, the crack would be loaded under a mixed-mode condition. In general, such R-curves are of great importance to characterise the crack growth behaviour, but their dependence on specimen geometry and stiffness was often indicated. Consequently, it would be difficult to consider R-curves as intrinsic material properties if the tests are not standardised. Under ENF mode II loading, the crack propagation is often too rapid to obtain the R-curve.

The objective of this study is to investigate the effects of moisture absorption on the delamination behaviour of woven glass/epoxy composites. Specimens at different moisture contents were studied through DCB, ENF and SLB testing to obtain pure mode I, pure mode II and mixed-mode I + II fracture resistances,

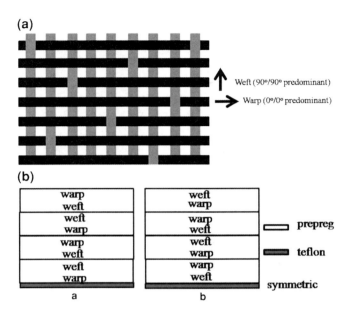

Fig. 1 a Schematic diagram of 8/8 harness satin-weave fabric; b Lay-up configuration for a 0°/0° and b 90°/90° predominant interface

respectively. SLB test has the advantage over mixed-mode bending (MMB) test that fracture toughness can be obtained directly from the experimental data using compliance calibration method [5]. In order to investigate the effects of the orientation of adjacent fibres on the interlaminar fracture behaviour, two series of specimens were prepared with 0°/0° or 90°/90° predominant fibre interface along the plane of the initial crack which was introduced at mid-thickness of specimens. The discussion concerned is not only on the fracture toughness for the crack initiation, but also on the R-curve under DCB loading for the crack propagation.

2 Experimental Details

The material used in this study was E-glass balanced fabric (8/8 h satin-weave) impregnated with XE85 epoxy (Ref: 1037/1250EP/XE85AI/35%, Hexcel). Laminates were fabricated from 8 prepreg layers cured at 120°C for 2 h at pressure of 3 bars (300 kPa). The thickness of the laminates was about 3 mm and the percentage of fibres was approximately 65% by weight. To initiate the pre-crack, a layer of Teflon film at thickness of approximately 15 μm was inserted at mid-thickness during the lay-up. In an eight-harness satin weave fabric, if warp strands are the predominant fibres on one face, then weft strands will be on another face (Fig. 1a).

Fig. 2 Specimen test configurations of **a** DCB, **b** SLB, and **c** ENF tests

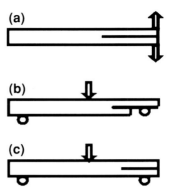

In order to study the effects of predominant fibre orientations of adjacent plies on the fracture toughness, two different types of laminates were fabricated. For the first type, the initial crack would propagate between 0°/0° predominant fibre interface; and for the second type, the crack propagation would be between 90°/90° (Fig. 1b). These specific lay-ups lead to symmetric configuration not only for the whole laminate, but also for each arm separated by the initial crack.

After fabrication was completed, the laminates were cut into testing coupons at dimensions of $100 \times 20 \times 3$ mm^3 by using diamond-coated abrasive blade.

For moisture absorption test, before being immersed into distilled water at atmospheric pressure and at the temperature of 72°C, laminates were dried in an oven at 70°C for 24 h to obtain the dry weight. Regularly, the specimens were taken out and wiped to remove the excessive water from the surface. After that, weight gain of the laminates was carefully measured over the testing period as a function of immersion time using an electronic balance with accuracy up to 1 mg. There were three immersion periods: 0 (dry), 60 and 118 days. The moisture content absorbed by the specimen, M is expressed as:

$$M = \frac{\text{weight of specimen } - \text{ weight of dry specimen}}{\text{weight of dry specimen}} \times 100\% \qquad (1)$$

At each level of moisture exposure period, the specimens were tested under double cantilever beam (DCB mode I), single leg bending (SLB mode I + II) and end notched flexural (ENF mode II) loadings (Fig. 2). All tests were performed at a constant crosshead speed of 1–3 mm/min at room temperature. For each moisture content and test configuration, at least six specimens of different initial crack length were loaded up to the crack extension. The measurement of the slope of load–displacement curve of each specimen provides the information necessary to calibrate the compliance as a function of the initial crack length experimentally.

2.1 Data Reduction Methods

2.1.1 Experimental Compliance Calibration

The delamination toughness in terms of the critical strain energy release rate (CSERR), G_C can be determined from the general relationship:

$$G_C = \frac{P_C^2}{2b} \frac{dC}{da}, \tag{2}$$

where P_C = critical load corresponding to crack initiation; b = width; C = compliance and a = initial crack length.

If the specimen maintains a constant fracture mode during crack propagation, the compliance, C can be calibrated experimentally as a function of the initial crack length. It can be done by measuring the slope of the linear part from the load–displacement curve, noted k, and the inverse of the slope gives the compliance for each specimen at different initial crack length. Then, the values of the compliance are interpolated as a function of crack length. Empirical compliance calibrated models proposed in the literature can be expressed by:

$$C_{\exp 1} = A + Ba^3, \tag{3}$$

$$C_{\exp 2} = \alpha a^n, \tag{4}$$

where A, B, α and n are constants to be determined empirically. Substituting Eq. 3 or 4 into Eq. 2 gives the following expressions:

$$G_{C\exp 1} = \frac{P_C^2}{2b} \cdot 3Ba^2, \tag{5}$$

$$G_{C\exp 2} = \frac{P_{C_f}^2}{2b} \cdot \alpha n a^{n-1}, \tag{6}$$

It is noted that Eq. 5 can be applied to the different fracture modes whereas Eq. 6 is only used for the DCB case. Figure 3 shows an example of the experimental calibration curves for DCB specimens.

These models were also applied to the crack propagation so as to determine a_p, defined as effective crack length as the crack grows and G_P, the strain energy release rate corresponding as following:

$$a_p = \left(\frac{C_p - A}{B}\right)^{\frac{1}{3}}, \; G_p = \frac{P_p^2}{2b} \cdot 3 \cdot B \cdot a_p^2 \tag{7}$$

$$a_p = \left(\frac{C_p}{\alpha}\right)^{\frac{1}{n}}, \; G_F = \frac{P_p^2}{2b} \cdot \alpha \cdot n \cdot a_p^{n-1} \tag{8}$$

where C_p and P_p signify the measured specimen compliance and the load corresponding to an effective crack length a_F.

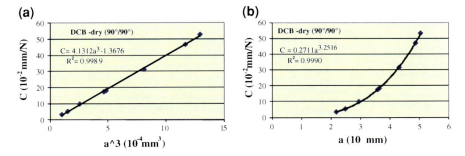

Fig. 3 Interpolation of the measured compliances obtained from the DCB-dry (90°/90°) specimens using **a** Eq. 3, and **b** Eq. 4

Then R-curves can be obtained in order to investigate the behaviour of crack propagation. In our work, the increment of the resistance relative to the fracture toughness, defined as $dR = G_P - G_{IC}$, will be plotted as a function of the extension of the crack: $da = a_P - a$, where a is initial crack length.

2.1.2 Beam Theory

In the case of DCB pure mode I tests, the compliance model based on the simple beam can be written as

$$C = \frac{8a^3}{Ebh^3}, \qquad (9)$$

where h is the thickness of one arm of the DCB specimen and E is the apparent Young's modulus. The determination of the value of E at a specified moisture level is based on Eq. 9 and the measurement of specimen stiffness, k as described above:

$$E = \frac{8a^3}{bh^3} \times k, \qquad (10)$$

By expressing the derivative dC/da in terms of C, and substituting into Eq. 2, the following equation is obtained:

$$G_{IC} = \frac{12 P_C^2 a^2}{E b^2 h^3} = \frac{3 P_C^2}{2abk}, \qquad (11)$$

During stable crack propagation, this model conduces to following expressions for determining a_p and G_P, which are defined as effective crack length and the strain energy release rate during the crack propagation, respectively:

$$a_p = \left(\frac{Ebh^3 C_p}{8}\right)^{\frac{1}{3}}, \; G_p = \frac{12 P_p^2}{E b^2 h^3} a_p^2 \qquad (12)$$

For ENF pure mode II tests, in the same manner, the expressions for E and G_{IIC} can be calculated as:

$$E = \frac{3a^3 + 2L^3}{8bh^3} \times k \tag{13}$$

$$G_{IIC} = \frac{9P_C^2 a^2}{16Eb^2 h^3} = \frac{9P_C^2 a^2}{2b(3a^3 + 2L^3)k} \tag{14}$$

where L is the half span length of the specimen.

In fact, the loading of SLB specimens can be considered as a combination of symmetrical and anti-symmetrical ones corresponding to mode I (DCB) and mode (ENF) cases respectively, where the mixed-mode ratio is estimated to be $G_{II}/G_T = 3/7$. The expressions used to determine E and the total critical fracture toughness, G_{TC} can be reduced as

$$E = \frac{7a^3 + 2L^3}{8bh^3} \cdot k \tag{15}$$

$$G_{IC} = \frac{3P_C^2 a^2}{4Eb^2 h^3}, G_{IIC} = \frac{9P_C^2 a^2}{16Eb^2 h^3} \quad \text{and} \quad G_{TC} = G_{IC} + G_{IIC} = \frac{21P_C^2 a^2}{16Eb^2 h^3} \tag{16}$$

3 Results and Discussion

3.1 Moisture Absorption

Figure 4 shows the change in moisture content as a function of the square root of immersion time. One-dimensional diffusion process is assumed because the in-plane dimension of specimens is much more important than the thickness. According to Fick's law [17]:

$$M = M_m \left\{ 1 - \frac{8}{\pi^2} \sum_{j=0}^{\infty} \frac{\exp\left[-(2j+1)^2 \pi^2 \left(\frac{Dt}{h^2}\right)\right]}{(2j+1)^2} \right\} \tag{17}$$

The diffusion coefficient of the composite, D is determined experimentally as 2.24×10^{-8} mm^2/s with the maximum moisture content, $M_m = 3\%$ by the following equation

$$D = \pi \left(\frac{h}{4M_m}\right)^2 \left(\frac{M_2 - M_1}{\sqrt{t_2} - \sqrt{t_1}}\right)^2 \tag{18}$$

Besides, Fig. 4 depicts a good fit of the experimental data with the Fickian diffusion curve in the linear region, which confirms the applicability of the Fickian model on these materials. Also, the moisture absorption is insensitive to fibre orientation, which conforms the finding by Candido et al. as well [14].

Fig. 4 Variation in the moisture content with square root of time of glass woven fibre/epoxy composite specimens immersed in distilled water at 72°C

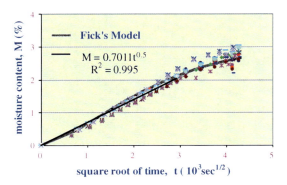

From Fig. 4, it is observed that the moisture content increases linearly at the initial stage. This suggests that the water build-up is diffusion controlled. According to some researchers, water is penetrated into the hydrophilic sites of the matrix as the consequence of capillarity due to the difference in the moisture concentration [4, 18]. In addition, moisture could be absorbed through microchannels network along the imperfect interface, as proposed by Tsenoglou et al. [19]. According to Cunha et al. [20], microcracks and voids are believed to contribute in moisture transport as well. Beyond the linear region, the increment rate decreases gradually. Costa [21] suggests that this is the pseudo-equilibrium state, where polymeric chain relaxation is believed to have started. In this study, complete saturation is not yet reached. Extension in moisture exposure period is needed if necessary.

3.2 Effects of Moisture Content on Flexural Modulus

For all three test configurations, the apparent flexural modulus can be determined from the measurement of the stiffness of each specimen as described above.

Figure 5 represents experimental load–displacement curves for three ENF specimens having the same dimension at different moisture content. It is evident that the stiffness of the specimens as well as the critical load at which the crack extension is initiated decrease with the moisture content.

Table 1 presents the average loss of the apparent flexural modulus relative to dry specimens due to moisture exposure. Results reveal that in general the apparent flexural modulus decreases with the increment in moisture content except for the cases of 90°/90° specimens under SLB and ENF testing. The high moisture content results in more than 20% loss in modulus; however, this loss is not linearly related to the moisture content. Reduction in tensile, compressive and flexural modulus due to moisture exposure was also reported elsewhere [18, 20, 22, 23].

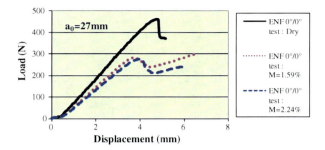

Fig. 5 Experimental load–displacement curves for ENF 0°/0° test

Table 1 Normalised flexural modulus obtained from DCB, SLB and ENF tests

	DCB		SLB		ENF	
E/E_{dry}	0°/0°	90°/90°	0°/0°	90°/90°	0°/0°	90°/90°
M = 1.59%	0.87	0.76	0.90	0.78	0.89	0.77
M = 2.24%	0.71	0.68	0.73	0.82	0.75	0.85

Degradation in the flexural modulus is believed to be mainly contributed by matrix and fibre/matrix interface degradation. As proposed by some researchers, hydrolysis separates the unsaturated groups in the polymer chains apart and weakens the interface [24–27]. Besides, in the work of Tompkins [28], it is indicated that as the outer most layer of the composite is directly exposed to moisture, swelling occurs. However, the inner layers which are still dry impose constraint to the swelling effect. Hence, additional hygroscopic stress is induced. Biro et al. [29] also suggested that matrix swelling could have released the radial pressure partially, thus weakening the interfacial bonding. Also, as moisture pickup rate in the epoxy is higher than the fibre, moisture mismatch occurs. This leads to different volumetric expansion between both constituents, hence causing variation in the local stress, as discussed by Mula et al. [30]. These additional dilatational stresses do not favour the mechanical performance of the composites. Internal stress could stretch the fibres and lead to high level of straightening, which endanger the entire composite. Kalfon et al. [18] mentioned that the stress variation due to moisture gradient could subsequently lead to delamination due to micro-damage formation.

In addition, the loss in modulus after immersion could be attributed to matrix plasticisation as well. As water which acts as plasticiser penetrated into the matrix, the ductility of the matrix is enhanced. Consequently, the matrix becomes softer and hence the modulus decreases, which is as discussed by Kalfon et al. [20].

Table 2 Critical energy release rate obtained from DCB tests

DCB test		0°/0°			90°/90°	
G_{IC} (J/m^2)	EXP1	EXP2	BT	EXP1	EXP2	BT
Dry	307.5	315.4	292.5	407.0	399.8	368.8
M = 1.59%	231.4	242.0	253.4	355.8	333.9	320.2
M = 2.24%	255.3	267.6	258.5	288.1	283.8	274.1

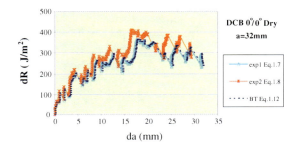

Fig. 6 R-curves obtained from DCB tests on dry specimens with 0°/0° interface

3.3 Comparison of Data Reduction Models

Table 2 displays the G$_{IC}$ results for all DCB specimens. G_{ICexp1}, G_{ICexp2} and G_{IC-BT} are obtained by applying Eqs. 5, 6 and 11 respectively. A good agreement between them is shown and the maximum difference is within ±10%. Hence, the compliance calibration can be done by choosing any one of them.

Regarding the crack propagation, Fig. 6 gives an example of R-curves obtained by three models Eqs. 7, 8 and 12 for the dry 0°/0° specimen with the initial crack length of 32 mm. Although there is no significant difference between theses data reduction models, the results obtained by Eq. 7 and Eq. 12 look very close but lower than those obtained by Eq. 18. This observation is true for all of the DCB specimens tested.

The same conclusion is drawn for SLB and ENF specimens by comparing G_{Cexp1} and G_{C-BT}. Since the experimental calibration described by Eqs. 5 and 7 take into account the experimental factors and is universal form for all tests, only the results obtained from Eq. 5 and Eq. 7 are used for discussion in the following.

3.4 Effects of the Predominant Fibre Orientation of Adjacent Plies on Fracture Toughness

The results of fracture toughness obtained from the specimens with crack growth between 0°/0° and 90°/90° fibre interfaces are compared in Fig. 7.

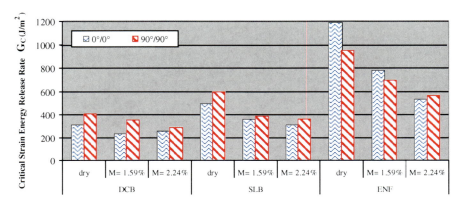

Fig. 7 Effect of the fibre orientation of adjacent plies to crack growth plane on fracture toughness

It is shown that under DCB pure mode I and SLB mixed-mode I + II loadings, the resistance to crack initiation of 90°/90° specimens is better than that of 0°/0° specimens at any moisture content. The difference is more considerable in DCB tests. However, an inverse influence of predominant fibre orientation is observed for pure mode II case except at the moisture content of 2.24%. In fact, the effect of fibre orientation is believed relative to the interface behaviour. The results obtained on 0°/0° interface specimens reveal that the shear strength of interface seems more sensitive to the interfacial degradation caused by water absorption than peel strength. If the crack propagates between 0° fibre interfaces is considered dominated by peel strength under DCB loading, by shear strength under ENF loading, but the crack growth between 90° fibre interfaces could be more complicated, because the interfacial cracking around the fibres can be developed much more in the circumferential direction. Herein, the shear strength of interface plays a role even under DCB loading, and the peel strength influence the fracture toughness under ENF loading as well. In reality, the fracture mode is rather mixed at the interface between fibres and the matrix. This phenomena can explain why the toughness ratio of pure mode II over pure mode I, in terms of G_{IIC}/G_{IC}, is more important in 0°/0° specimens than that of 90°/90° ones whatever the moisture content M (Fig. 8). This difference decreases with increasing M and it disappears when $M = 2.24\%$.

3.5 Effects of the Moisture Content on the Fracture Toughness

Table 3 illustrates the loss of fracture toughness due to moisture absorption relative to dry specimens, which is found significant. In general, the value of G_C decreases with the increment of the moisture content regardless of the loading

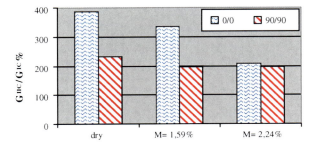

Fig. 8 Variation of the toughness ratio: G_{IIC}/G_{IC} with M and different fibre orientations

Table 3 Normalised fracture toughness G_C/G_{Cdry} obtained from DCB, SLB and ENF tests

	DCB		SLB		ENF	
G_C/G_{Cdry}	0°/0°	90°/90°	0°/0°	90°/90°	0°/0°	90°/90°
M = 1.59%	0.75	0.87	0.73	0.65	0.65	0.73
M = 2.24%	0.83	0.71	0.63	0.60	0.45	0.60

mode and the predominant fibre interface except DCB 0°/0° specimens. The general observation is that for both types of predominant fibre orientations, more important the contribution of mode II is, more significant the effect of moisture absorption on the fracture resistance. In the case of crack growth between 0°/0° fibre interfaces, the reduction in fracture toughness attained to 55% under ENF pure mode II loading.

The reduction in fracture toughness due to moisture absorption is believed to be caused by matrix or/and interfacial degradation as discussed previously. This effect was also reported by Scida et al. [32]. The increase of G_C observed in the work of Buehler and Seferis [33] could be due to plasticisation by water that enhances the matrix ductility. Furthermore, some authors reported that as the immersion period is longer, more voids are formed and the matrix becomes more porous. Consequently, crack tip blunting occurs, where the crack tip radius is increased and stress concentration at the crack tip is partially released [5, 34]. Hence, this favours the mode I fracture toughness. However, in our study, it is noted that even the G_{IC} at $M = 2.24\%$ is still lower than the dry one, which draws the general conclusion that moisture is harmful to the delamination behaviour of the composite.

Otherwise, the toughness ratio: G_{IIC}/G_{IC}, decreases with increasing moisture content M (Fig. 8). It is well known that this ratio depends strongly on the brittleness of matrix: G_{IIC}/G_{IC} can attain more than twenty in the case of brittle matrix, whereas it can be equal to 1 if matrix is very ductile. Hence, it seems that the water acts as plasticiser on the matrix, which becomes more ductile [18].

Fig. 9 The variation of the G_{TC} value as a function of mixed-mode ratio

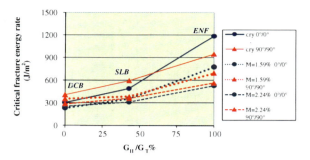

3.6 Variation of Fracture Toughness as a Function of Mixed-Mode Ratio

In Fig. 9, the values of fracture toughness are plotted as a function of the mixed-mode ratio, G_{II}/G_T. The increment of fracture toughness with the mixed-mode ratio is evident in all cases, which corresponds well to the general conclusions found in the literature [5, 35].

A mixed-mode criterion proposed by Gong et al. [35, 36] assumes that:

$$G_{TC} = G_{IC} + (G_{IIC} - G_{IC})\left(\frac{G_{II}}{G_I + G_{II}}\right)^m \qquad (19)$$

where m is the material constant and has to be determined empirically. This criterion was then applied by Benzeggagh and Kenane, named BK criterion in the literature [37]. Figure 10 compares the measured G_{TC} values for SLB specimens with those predicted by Eq. 19 using $m = 2/3$. A good agreement is shown for the majority of the cases. It is interesting to note that although the values of G_{IC} and G_{IIC} decrease with the moisture absorption, the material constant, m does not appear sensitive to moisture exposure.

3.7 R-curves Obtained from DCB Mode I Specimens

Under DCB loading, as an example, Fig. 11 compares the load–displacement curves obtained from the 0°/0° as well as 90°/90° specimens at different moisture contents. Here the specimens have the initial crack length about 23 mm and the departure of displacements of the specimens is shifted of 5 mm one from another to give a clearer view. It is seen that whatever the moisture content, the resistance to delamination is better in 90°/90° specimens than 0°/0° ones. Otherwise, all curves evolve in nearly the same way: at the beginning, the force increases linearly until the critical load, where the crack initiation occurs; and then the crack grows slowly as load increases up to the peak load, herein the first local instable crack

Fig. 10 Comparison between the measured G_{TC} values and the predicted values by the criterion with $m = 2/3$ for SLB specimens

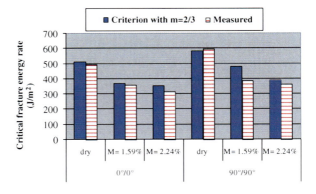

Fig. 11 Load–displacement curves of DCB tests at different moisture content and predominant fibre orientation

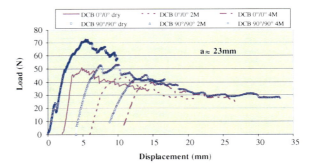

propagation is observed accompanied by a load drop. After that, the load decreases progressively as the crack propagates in stick–slip manner for a long time.

The R-curve describes the variation of the resistance to delamination as a function of the crack extension. As an example, Fig. 12a presents the crack behaviour of the wet specimens (4 months immersion in water) with 0° predominant fibres at interface. The increment of the resistance to crack growth for each specimen is normalised by its own fracture toughness so as to eliminate the scatter of G_{IC} between tests. All specimens are considered the same geometry except of the initial crack length. It is shown that the R-curves are indeed independent of the initial crack length by taking into account the dispersion of the results. For a composite system, the dispersion of experimental results is usually large and inevitable due to many defects within the material. Even for two specimens with the same dimensions, their normalised R-curves can be quite different (Fig. 12b).

It is seen that the resistance to delamination growth for all specimens tested increases steeply and linearly with the crack extension in the first time. This phenomenon results in damage mechanisms in the zone around the crack tip such as plastic deformation in resin; fibre bridging and breaking; multiple cracking and crack shifting. Then the rate of increment in resistance becomes slower and slower with unstable stick–slip crack , which can be explained by the development of the

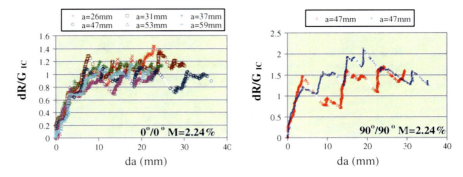

Fig. 12 Normalised R-curves of wet DCB specimens (4 months immersion in water M = 2.24%)

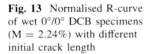

Fig. 13 Normalised R-curve of wet 0°/0° DCB specimens (M = 2.24%) with different initial crack length

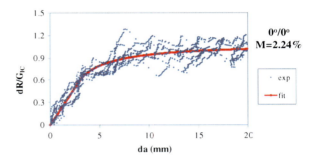

large scale bridging and its breakage [16] and the bridging fibres could be interlocked and then broken or peeled at the weft/warp intersections. As soon as the formation of fibre bridging creates a balance with the fibre breaking, the resistance to crack growth attains a stable level, noted G_{IR}.

In order to obtain a representative R-curve for a testing configuration, all results have to be taken into account. It seems that the normalised R-curves can be divided into linear and nonlinear regions (Figs. 5 and 12). Hence, each region has to be modelled by using different equation. We propose to use a linear equation for the first region and a non-linear equation for the second region:

$$\frac{dR}{G_{IC}} = \alpha_1 + \beta_1(da) \tag{20a}$$

$$\frac{dR}{G_{IC}} = \alpha_2 + \beta_2 \frac{1}{da} \tag{20b}$$

where α_i and β_i are the constants fitted from the experimental results. The results show good correlations between the experimental data and the results obtained from the proposed models (Fig. 13).

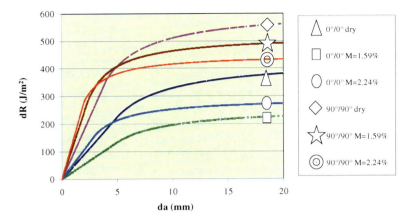

Fig. 14 R-curves measured on 0°/0° and 90°/90°DCB specimens at different moisture content

In this manner, the experimental results of at least three specimens with different initial crack lengths have been exploited to fit each normalised R-curve. Then, all R-curves can be obtained by multiplying the fitted normalised R-curves by its own fracture toughness, respectively. Figure 14 compares all of the R-curves measured on 0°/0° and 90°/90° DCB specimens at different moisture contents.

It is revealed that at all moisture content the resistance to delamination propagation, in terms of $(G_p - G_C)$, is always better in the 90°/90° specimens than 0°/0° ones. In fact, the mechanism of crack propagation between 0° or 90° fibre interfaces is quite different.

In the specimens with crack propagation between 90° predominant fibres, a lot of weft strands peeled off from the edges of the specimen. This phenomenon leads to the rapid formation of fibre bridging in the transverse direction so as to provide greater slope of the linear part of the R-curve, which represents the resistance increment per unit of crack extension. The higher the moisture content is, more degradation in the interface, the greater the slope. Regarding the maximum of the resistance to delamination growth at stable level, named G_{IR}, at 90°/90° interface, the effect of moisture content is inversed. The value of G_{IR} decreases with the increasing of the moisture content. That means the quantity of the fibre bridging at the stable level of the resistance could become smaller in wet specimens than that in dry specimens.

In the cases of the crack growth between 0° predominant fibres, the fibre bridging in the longitudinal direction at the crack tip looked more extended whereas less fibre peeling from the edges of the specimens was observed. The slope and the value of G_{IR} are much lower than those with 90° fibre interface. However, their variation as a function of the moisture content follows the same tendency than that of 90°/90° specimens except the case of the specimens at $M = 1.59\%$. In fact, as great dispersion of the R-curves obtained from these specimens was found, only the results obtained from three specimens have been used to fit its R-curve.

Our observation is in good agreement with that of Gill et al. [31] on woven carbon/epoxy composite. It is not only reported that the G_{IC} value obtained from the specimens with 0°/90° interface is 60% higher than that from 0°/0° specimens, but the interlocking of the bridging fibres at the warp/weft intersections during the crack extension was observed as well. This phenomenon resulted in the retardation of the crack growth.

In conclusion, the degradation of fibre-matrix interface caused by water absorption accelerates the formation of fibre bridging beyond the crack tip in the early crack growth, hence leads to higher rate of increment in the resistance. However, the quantity of the fibre bridging decreases with the moisture content at the stable level of the resistance to delamination. The increment of the resistance to delamination growth is always higher in the 90°/90° specimens than that in the 0°/0° ones at all moisture contents.

4 Conclusions

Interlaminar fracture toughness in mode I, mixed-mode I + II and mode II of woven glass/epoxy composites subjected to different moisture absorption were examined experimentally. Based on the results obtained, it can be concluded that:

1. The moisture absorption of 8/8 h satin weave glass/epoxy composites can be well described by Fick's law.
2. The G_{TC} values obtained from the models based on simple beam theory and experimental compliance calibration are found to be similar.
3. The apparent flexural modulus: E, and the fracture toughness: G_C decrease with the increment of moisture content. The maximum reduction of E and G_C is found to be about 20% and 50%, respectively.
4. For both of 90°/90° and 0°/0° specimens, the influence of moisture absorption on the delamination behaviour of the composites becomes more significant if the participation of mode II is more important. Moreover, the toughness ratio: G_{IIC}/G_{IC}, decreases with increasing moisture content, its reduction is more remarkable if the crack grows between 0°/0° interface than that between 90°/90° interface.
5. Under pure mode I and mixed-mode I + II loadings, the fracture toughness measured at 90°/90° interface is greater than that at 0°/0° interface. However, an inverse effect is shown in pure mode II case.
6. A mixed-mode criterion described by the equation :$G_{TC} = G_{IC} + (G_{IIC} - G_{IC}) \left(\frac{G_{II}}{G_I+G_{II}}\right)^m$ with $m = 2/3$ agrees well with the experimental results regardless the moisture content and the adjacent fibre direction. It means that although the

values of G_{IC} and G_{IIC} decrease with the moisture content, the material constant m seems to be independent of water absorption.
7. Regarding the behaviour of the crack growth, the increment of the resistance to delamination growth is always higher in the 90°/90° specimens than that in the 0°/0° ones at all moisture contents. The degradation of fibre/matrix interface caused by water absorption accelerates the formation of fibre bridging beyond the crack tip in the early crack growth, hence leads to higher rate of increment in the resistance. However, the quantity of the fibre bridging decreases with the moisture content at the stable level of the resistance to delamination.

References

1. Daniel, I.M., Ishai, O.: Engineering mechanics of composite materials. 2nd edn., Oxford University Press, USA (2005)
2. Zenasni, R., Bachir, A.S., Vina, I., et al.: Effect of hygrothermomechanical aging on the interlaminar fracture behaviour of woven fabric fibre/PEI composite materials. J. Thermoplast Compos. **19**, 385–398 (2006)
3. Huang, G., Sun, H.: Delamination behaviour of glass/polyester composites after water absorption. Mater Design **29**, 262–264 (2008)
4. Davidson, B.D., Kumar, M., Soffa, M.A.: Influence of mode ratio and hygrothermal condition on the delamination toughness of a thermoplastic particulate interlayered carbon/epoxy composite. Compos. Part A-Appl. Sci. **40**, 67–79 (2009)
5. Walker, L., Hu, X.Z.: Mode I delamination behaviour of short fibre reinforced carbon fibre/epoxy composites following environmental conditioning. Compos. Sci. Technol. **63**, 531–537 (2003)
6. Marom, G., Roman, I., Harel, H., Rosensaft, M.: The characterization of mode I mode II delamination failures in fabric-reinforced laminates. Proceedings of the ICCM6 & ECCM3, **3**, 215–273 London (1987)
7. Chai, H.: The characterization of mode I delamination failure in non-woven multidirectional laminates. Composite. **15**, 277–290 (1984)
8. Nicholls, D.J., Gallagher, J.P.: Determination of G_{IC} in angle ply composite using a cantilever beam test method. J. Reinf. Plast. Comp. **2**, 2–17 (1983)
9. Benzeggagh, M.L., Gong, X.J., Laksimi, A., Roelandt, J.M.: On the test of mode I delamination and the importance of stratification. Polym. Eng. Sci. **31**, 1286–1292 (1991)
10. Gilchrist, M.D., Svensson, N.: A fractographic analysis of delamination within multidirectional carbon/epoxy laminates. Compos. Sci. Technol. **55**, 195–207 (1995)
11. Ozdil, F., Carlsson, L.A.: Beam analysis of angle-ply laminate DCB specimens. Compos. Sci. Technol. **59**, 305–315 (1999)
12. Chou, I., Kimpara, I., Kageyyama, K., Ohsawa, I.: Mode I and mode II fracture toughness measured between differently oriented plies in graphite/epoxy composites. In: Composite Materials. Fatigue and Fracture-Fifth Volume, ASTM STP **1230**,132–151 (1995)
13. Gong, X.J., Hurez, A., Verchery, G.: On the determination of delamination toughness by using multidirectional DCB specimens. Polym. Test. **29**, 658–666 (2010)
14. Candido, G.M., Costa, M.L., Rezende, M.C., et al.: Hygrothermal effects on quasi-isotropic carbon epoxy laminates with machined and molded edges. Compos. Part. B-Eng. **39**, 490–496 (2008)
15. Gong, X.J., Vannucci, P., Verchery, G.: Effects of sdjacent fibre direction on the resistance of laminates to delamination fracture. Pap1293, Proceedings of ICCM13, Beijing, China, June (2001)

16. Keiichiro, T., Yutaka, H., Hitoshi, I., Kazunori, S.: Mode I interlaminar fracture toughness and fracture mechanism of angle-ply Carbon/Nylon laminates. J. Compos. Mater. **30**(6), 651-661 (1996)
17. Shen, C.H., Springer, G.S.: Moisture absorption and desorption of composite materials. J. Compos. Mater. **10**, 2–20 (1976)
18. Kalfon, E., Harel, H., Marom, G., et al.: Delamination of laminated composites under the combined effect of nonuniform heating and absorbed moisture. Polym. Compos. **26**, 770–777 (2005)
19. Tsenoglou, C.J., Pavlidou, S., Papaspyrides, C.D.: Evaluation of interfacial relaxation due to water absorption in fibre-polymer composites. Compos. Sci. Technol. **66**, 2855–2864 (2006)
20. Cunha, J.A.P., Costa, M.L., Rezende, M.C.: Study of the hygrothermal effects on the compression strength of carbon tape/epoxy composites. Lat. Am. J. Solids Stru. **5**, 157–170 (2008)
21. Costa, M.L.: Efeito do conteúdo de vazios no comportamento mecânico de compósitos avançados carbono/epóxie carbono/bismaleimida. Ph.D. thesis, Instituto Tecnológico de Aeronáutica, São José dos Campos (2002)
22. Abdel-Magid, B., Zaiee, S., Gass, K., et al.: The combined effects of load, moisture and temperature on the properties of E-glass/epox composites. Compos. Struct. **71**, 320–326 (2005)
23. Aktas, L., Hamidi, Y., Cengiz Altan, M.: Effect of moisture on the mechanical properties of resin transfer molded composites–Part I: Absorption. J. Mater. Process. Manu. **10**, 239–254 (2002)
24. Aoki, Y., Yamada, K., Ishikawa, T.: Effect of hygrothermal condition on compression after impact strength of CFRP laminates. Compos. Sci. Technol. **68**, 1376–1383 (2008)
25. Apicella, A., Migliaresi, C., Nicolais, L., et al.: The water ageing of unsaturated polyester-based composites; influence of resin chemical structure. Composite **14**, 387–392 (1983)
26. Bradly, W.L., Grant, T.S.: The effect of the moisture absorption on the interfacial strength of polymeric matrix composites. J. Mater. Sci. **30**,5537–5542 (1995)
27. Gellet, E.P., Turley, D.M.: Seawater immersion ageing of glass-fibre reinforced polymer laminates for marine applications. Composite **30A**, 1259–1261 (1999)
28. Tompkins, S.S., Tenney, D.R., Unna, J.: Prediction of moisture and temperature changes in composites during atmospheric exposure. In: Composite Materials. Testing and Design-Fifth conference, ASTM STP Vol. **674**,368–382 (1979)
29. Biro, D.A., Pleizier, G., Deslandes, Y.: Application of the microbond technique effects of hygrothermal exposure on carbon fibre/epoxy interfaces. Compos. Sci. Technol. **46**, 293–301 (1993)
30. Mula, S., Ray, B.C., Ray, P.K.: Assessment of interlaminar shear strength of hybrid composites subjected to a fluctuating humid environment. International Symposium of Research Students on Materials Science and Engineering, Chennai, India (2004)
31. Gill, A.F., Robinson, P., Pinho, S., Sargent, J.: Effect of RTM defects on mode I and II delamination behaviour of 5HS woven composites. 16th international conference on composite materials, Kyoto, Japan (2007)
32. Scida, D., Aboura, Z., Benzeggagh, M.L.: The effect of ageing on the damage events in woven-fibre composite materials under different loading conditions. Compos. Sci. Technol. **62**, 551–557 (2002)
33. Buehler, F.U., Seferis, J.C.: Effect of reinforcement and solvent content on moisture absorption in epoxy composite materials. Compos. Part A-Appl. Sci. **31**, 741–749 (2000)
34. Schaffer, J.P., Saxena, A., Antolovich, S.D., et al.: The science and design of engineering materials. Irwin, Chicago (1995)
35. Gong, X.J.: Rupture interlaminaire en mode mixte I + II de composites tarifiés unidirectionnels et multidirectionnels verre/epoxy. Ph.D. thesis, Université de Technologie de Compiègne, France (1992)

36. Gong, X.J., Benzeggagh, M.L.: Mixed-mode interlaminar fracture toughness of unidirectional glass/epoxy composite. In: Composite Materials, Fatigue and Fracture-Fifth Volume, ASTM STP **1230**, 100–123 (1995)
37. Benzeggagh, M.L., Kenane, M.: Measurement of mixed-mode delamination fracture toughness of unidirectional glass/epoxy composites with mixed-mode bending apparatus. Compos. Sci. Technol. **56**, 439–449 (1996)

The Study of Response of High Performance Fiber-Reinforced Composites to Impact Loading

O. Saligheh, R. Eslami Farsani, R. Khajavi and M. Forouharshad

Abstract There are numerous applications for high impact resistant fiber-reinforced composite especially in aerospace, automotive, marine, and military fields. Impact damages of foreign objects on composite structures is of great importance, because may severely reduce their strength and stability. These impacts are ranging from low speed impacts like dropping of a tool to high speed impacts when small sands or debris throw away by aircraft tires. This research focuses on fiber reinforced composites, impact tests, failure modes and damage propagation through the composite during destructive impact load.

Keywords Fiber-reinforced composite · Low- and high-velocity impacts · Failure modes

1 Introduction

During the life of a structure, impacts by foreign objects can be expected to occur during manufacturing, service, and maintenance operations. An example of impact occurs during aircraft takeoffs and landings, when stones and other small

O. Saligheh (✉)
Department of Textile Engineering, Islamic Azad University,
South Tehran Branch, Young Researchers Club, Tehran, Iran
e-mail: omid.saligheh@gmail.com

R. Eslami Farsani
Faculty of Mechanical Engineering, K. N. Toosi University of Technology,
Tehran, Iran

R. Khajavi · M. Forouharshad
Department of Textile Engineering, Islamic Azad University,
South Tehran Branch, Tehran, Iran

debris from the runway are propelled at high velocities by the tires. During the manufacturing process or during maintenance, tools can be dropped on the structure or in situations such as automobile accidents. In this cases impact velocity are small but the mass of the impactor is larger. Fiber-reinforced composite structures are more susceptible to impact damage than a similar metallic structure. In composite structures, impacts create internal damage that often cannot be detected by visual inspection. This internal damage can cause severe reductions in strength and can grow under load. Therefore, the effects of foreign object impacts on composite structures must be understood, and proper measures should be taken in the design process for composites to account for these expected events.

2 Fiber-Reinforced Composites

A composite is a material composed of two or more substances having different physical characteristics and in which each substance retains its identity while contributing desirable properties to the whole. Composites consist of two phases; one is continuous and is called matrix, and the other is discontinuous and is in form of fibers or particulates. The discontinuous phase is well dispersed in the matrix and acts as a reinforcement or modifier, to improve and alter the matrix properties.

Matrix materials used in composites are typically ceramics, metals, or polymers. Fibers are the common reinforcement due to their effectiveness, although particulates of various geometries are also used. Polymer matrix composites, with fibers reinforcement, are the most commonly used high performance composites and are widely used in many applications because of their unique properties and characteristics. High performance fiber-reinforced polymeric composites consist of high strength and modulus fibers embedded in or bonded to a polymeric matrix with distinct interfaces between them [1]. Fiber-reinforced polymeric composites have many distinctive advantages over traditional materials; the most significant one being their light weight. Low density leads to high specific strength and specific stiffness. Another benefit of polymer composites is their design flexibility and freedom. Table 1 shows advantages/disadvantages of advance composite [2].

Generally, the advantages accrue for any fiber reinforced composite combination and disadvantages are more obvious with some. These advantages have now resulted in many more reasons for composite use as shown in a nutshell in Table 2 [2]. Proper design and material selection can circumvent many of the disadvantages.

By varying the fibers type, the fibers direction and volume fraction, and adopting different manufacturing processes, the composites' properties and characteristics can be easily controlled and adjusted.

Matrices used in fiber-reinforced polymeric composites are either thermoset or thermoplastic polymers. They play several important roles in the composites,

Table 1 Advantages/disadvantages of advanced composites

Advantages	Disadvantages
Weight reduction; High strength and modulus to weight ratio	Cost of raw materials and fabrication
Tailorable properties; Can tailor strength or stiffness to be in the load direction	Transverse properties may be weak
Redundant load paths (fiber to fiber)	Matrix is weak, low toughness
Longer life (no corrosion)	Reuse and disposal may be difficult
Lower manufacturing costs because of less part count	Difficult to attach
Inherent damping	Analysis is difficult
Increased or decreased thermal or electrical conductivity	Matrix subject to environmental degradation

Table 2 The reasons for using composites

Reason for use	Composite material	Application
Lighter, stiffer, stronger	Boron, carbon/graphite, aramid	Military aircraft, better performance commercial aircraft
Controlled or zero thermal expansion	High modulus carbon/graphite	Spacecraft with high positional accuracy requirements for optical sensors
Environmental resistance	Fiberglass, vinylesters, epoxy	Tanks and piping, corrosion resistance to industrial chemicals, crude oil, gasoline at elevated temperatures
Lightweight, damage tolerance	High strength carbon/graphite, fiberglass, (hybrids), epoxy	CNG tanks for 'green' cars, trucks and buses to reduce environmental pollution
Less fatigue	Carbon/graphite/epoxy	Tennis, squash and racquetball racquets
Transparency to radiation	Carbon/graphite-epoxy	X-ray tables
Crashworthiness	Carbon/graphite-epoxy	Racing cars
Water resistance	Fiberglass, polyester or isophthalic polyester, vinylester, epoxy resins	Commercial boats
Lightweight, high strength and modulus, high energy dissipation	Aramid And HPPE fibers, Keraton, PU resins	Bullet proof vests and Helmets, protection panels, Vehicle protection
Lightweight, high strength and modulus	Glass, carbon and aramid fibers reinforced composite	Footbridge, civil engineering, earthquake protection, cable cars
Low cost, insulating characteristics	Carbon/graphite, glass fiber/phenolic	Electrical/electronics energy systems/communications

such as holding the fibers in place, transferring stress between fibers, and protecting the fibers from adverse environment or mechanical abrasion. Traditionally the matrices of fiber-reinforced polymeric composite are thermoset polymers, such as polyesters, epoxies, vinyl esters and phenolics. It is related to their low viscosity and their wettability properties. In addition, thermoset matrices have high Tg, good thermal stability and chemical resistance, and also exhibit low creep and stress relaxation behavior. However, thermoset polymers have long fabrication time due to the slow crosslinking and solidifying processes. Also, thermoset matrix can be brittle and fail at low strain and as a result, resistance to impact can be poor.

The most important benefit of thermoplastic matrices is their high fracture resistance and impact strength, due to their high strain to failure. Thermoplastic matrices have short fabrication times, are thermoformable and can be easily repaired by welding or solvent bonding. Since thermoplastics matrices are not crosslinked polymers, they can be easily recycled and reused. However, their wettability to fibers is relatively poor because their high viscosities and there is no direct chemical bonding between matrix and fibers.

Fibers are the major component that occupies the largest volume fraction in fibers reinforced polymer composites. They are principal load-bearing constituents that dominate the characteristics and properties of the composites. It is important to select the proper fibers in order to meet the requirements for the final product. There are a large variety of commercially available fibers with different properties that can be used as reinforcements: glass, carbon, aramid, high performance polyethylene (HPPE), ceramic fibers, and metal fibers.

Glass fibers are the most common reinforcement due to their low cost, high tensile strength, high chemical resistance and excellent insulating properties. The disadvantage of glass fibers are their low tensile modulus, high density, low resistance to abrasion and fatigue.

Carbon fibers have the advantage of high specific tensile modulus and strength, high fatigue performance and low coefficient of thermal expansion. The disadvantages include low impact resistance.

The fibers begin as an organic fiber, rayon, polyacrylonitrile or pitch which is called the precursor. The precursor is then stretched, oxidized, carbonized and graphitized. There are many ways to produce these fibers, but the relative amount of exposure at temperatures from 2500 to 3000°C results in greater or less graphitization of the fiber. Higher degrees of graphitization usually result in a stiffer fiber (higher modulus) with greater electrical and thermal conductivities and usually higher cost [1, 2].

Ceramic fibers have attracted considerable interests as reinforcement for metal and ceramic matrix composite materials due to its excellent mechanical and physical properties. Applications of ceramic fibers are in gas turbines, both aeronautical and ground-based, heat exchangers, containment walls for fusion reactors, as well as uses for which no matrix is necessary such as candle filters for high temperature gas filtration [3].

Aramid fibers, such as Kevlar® and Twaron® family, have higher specific tensile strength and strain to failure than carbon fibers, which offers them good damage tolerance against impact. Kevlar fibers have good chemical and thermal stability, and poor compressive properties. Their outstanding potential derived mostly from the anisotropy of their superimposed substructures presenting pleated, crystalline, fibrillar and skin–core characteristics. Nowadays these fibers are used in various structural parts including reinforced plastics, ballistics, tires, ropes, cables, asbestos replacement, coated fabrics, and protective apparel.

High performance polyethylene (HPPE) fibers, such as Dyneema® and Spectra® family possesses extraordinary physical and mechanical properties, such as low specific weight, high specific modulus and strength, high impact resistance, low dielectric constant, good UV resistance, low moisture absorption, excellent vibration damping capability, low coefficient of friction, excellent abrasion, cut resistance and self-lubricating properties [4–10]. The disadvantage could be low melting temperature of polyethylene fibers. They can be produced by a gel-spinning process from ultra high molecular weight polyethylene (UHMWPE); typical draw ratios are between $30\times \sim 100\times$ and the high strength and modulus of the fibers is determined by the draw [11].

The fibers used in composites can adopt many forms; they can be used as continuous filaments or yarns, as well as discontinuous chopped fibers. Two and three dimensional fabrics can be made from continuous fibers using textile processing such as weaving, knitting, braiding and needle-punching. Fabrics are easy to handle and provide good control over the fibers orientation and placement. The alignment of fibers in composites can be unidirectional, bidirectional, multidirectional or random.

Fibers volume fraction is an important parameter that dictates the composites' properties. Fibers volume fraction should be as high as possible to maximize the strength of the composites. Another important aspect of fiber-reinforced composites is the interfacial adhesion between fibers and matrix; good bonding strength is needed for fibers to achieve good reinforcement. The influence of interfacial adhesion on composites properties is profound, compression strength, flexural strength, shear strength, and impact damage resistance increase with the level of adhesion; although, high velocity impact resistance decreases with increase in bonding strength. Relatively poor adhesion makes interfiber and interlaminar debonding easier, thus high impact energy is efficiently dissipated laterally, and resistance to penetration is enhanced [9].

3 Low- and High-Velocity Impact Test

Different test apparatus are used to simulate various types of impact. Two types of tests are used more by most investigators, although many details of the actual test apparatus may differ. Drop–weight testers are used to simulate low-velocity impacts typical of the tool-drop problem, in which a large object falls onto the

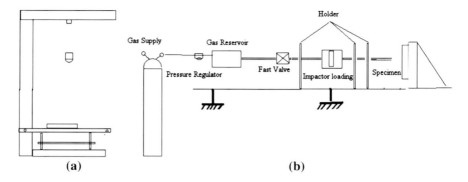

Fig. 1 The schematic of drop-weight (**a**), and gas-gun (**b**) systems

structure with low velocity. Gas-gun systems, in which a small impactor is propelled at high speeds, are used to simulate the type of impacts encountered during aircraft takeoffs and landings. Testing conditions should replicate the actual impacts the structure is designed to withstand, and so the choice of the proper test apparatus is important. The schematic of gas-gun and drop-weight systems are shown in Fig. 1.

In gas gun system high pressure compressed air or gas is drawn into an accumulator to a given pressure controlled by a regulator. The pressure is released by a solenoid valve, the breakage of a thin diaphragm, or other mechanism. The impactor then travels through the gas gun barrel and passes a speed-sensing device while still in the barrel of the gun or right at the exit and with different device the velocity of impactor is calculated [10–13]. Drop-weight systems are used extensively and can be of differing designs. Heavy impactors are usually guided by a rail during their free fall from a given height [14–17].

A large mass with low initial velocity may not cause the same amount of damage as a smaller mass with higher velocity, even if the kinetic energies are exactly the same.

Experimental studies attempt to replicate actual situations under controlled conditions. For example, the impact of a composite structure by a larger impactor at low velocity, which occurs when tools dropped accidentally on a structure or in automobile accidents. This situation is best simulated using a drop-weight system. Another concern is during aircraft take off and landing, debris flying from the runway can cause damage; this situation, with small high velocity impactor or projectile, is best simulated using a gas gun system.

4 Methods for Damage Assessment

Many techniques have been developed to determine the extent of impact-induced damage in composite structures. Impact damage is internal and generally consists of delaminations, matrix cracking, and fiber breakage, which usually cannot be detected by simply examining the surface of the specimen. In applications where one seeks to detect the presence of impact-induce damage, nondestructive techniques providing whole field information are used. For research purposes when the location of impact is known, other, destructive methods provide more detailed information but also end up destroying the specimen in the process [18].

4.1 Nondestructive Techniques

It is necessary to determine if damage is present, where it is located, and its extent. With translucent material systems such as glass/epoxy or Kevlar/epoxy composites, impact damage can be observed using strong backlighting. The size and shape of delaminations and the presence of matrix cracks can be detected by visual observation. Several investigators have used this technique, and much of the early work on impact damage was carried out using this technique and brought about a great deal of understanding about the morphology of impact damage and impact damage development.

Other material systems such as graphite/epoxy are opaque, and thus this visual inspection approach cannot be used. Whole-field nondestructive methods such as ultrasonic imaging or radiography are used to visualize internal damage over large area. C-scan and x-rays provide a projected image of the damage zone and are useful in delineating the extent of the damage, but many of the features of the damage area are lost. It is important to understand how delaminations are distributed through the thickness, their size and orientations, and how they might be connected through intraply cracks.

Many other nondestructive techniques for impact damage inspections have been reported in the literature such as infrared thermography, vibrothermography, high frequency low amplitude mechanical vibrations, and tapping technique, although ultrasonic and x-ray methods are by far the two most commonly used techniques [18].

4.2 Destructive Techniques

Detailed maps of image can be obtained by sectioning several strips of material at different locations and orientations throughout the impacted zone. After careful preparation, microscopic examinations of each section are used to construct detailed maps of delaminations at each interface and of matrix cracks in each ply; and also observation of damage development during impact by using high speed photography is another way for damage assessment [18].

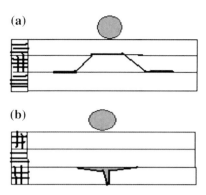

Fig. 2 Two types of matrix cracks: **a** tensile crack, **b** shear cracks

5 Failure Modes of Fiber-Reinforced Composites in Impact Damage

Numerous experimental and analytical investigations have been undertaken to uncover the penetration failure modes of the composites under impact [16–19].

There are many parameters that affect the initiation and growth of impact damage. Material properties affect the overall stiffness of the structure and the contact stiffness and therefore will have a significant effect on the dynamic response of the structure. Fibers, matrix, and fiber-matrix interface have an important role and control the initiation and growth of impact damage. The thickness of the laminate, the size of the composite, and the boundary conditions are all factors that influence the impact dynamics, since they control the stiffness of the composite. The characteristics of the impactor including its density, elastic properties, shape, initial velocity, and incidence angle are another set of parameters that should be considered. The effect of layup, stitching, and environmental conditions are important factors that have received various degrees of attention. For impacts that do not result in complete penetration of the target, experiments indicate that damage consists of delaminations, matrix cracking, and fiber failures. In general for low velocity impacts damage starts with the creation of a matrix crack. In some cases the target is flexible and the crack is created by tensile flexural stresses in the bottom ply of the laminate. This crack, which is usually perpendicular to the plane of the laminate, is known a tensile crack (Fig. 2a). For thick laminates, cracks appear near the top of the laminate and are created by the contact stresses. These cracks are known shear cracks (Fig. 2b). Matrix cracks induce delamination at interfaces between adjacent plies and initiate a pattern of damage evolution either from bottom up or from top down [18].

Delaminations, that is, the debonding between adjacent laminates, are of most concern since they significantly reduce the strength of the laminate.

In high velocity impacts failure during laminate penetration depends in a large part on the shape of the impactor, which strongly affects the perforation energy. In general, shear plugging occurs near the impacted side, followed by a region in which failure occurs by tensile fiber fracture, and near the exit, delamination occur. Depending on composite rigidity, impactor mass, and velocity, in some case only a shear failure mode is observed. In other case, the first two modes are present, and for thicker targets all three modes are present. The penetration energy can be estimated by adding the energies required to produce each type of failure involved in the penetration process: fiber failure, matrix cracking, delamination, and friction between the impactor and composite [16–19].

High-velocity impact is a highly dynamic event involving transient stress wave propagation. In many cases, the impacted material will fail before the stress waves reflect from the material boundaries. During the impact by a spherical impactor, the stress distribution under the impactor is truly three-dimensional. As soon as the impactors enter in contact with the target, a compressive wave, a shear wave, and surface waves propagate away from the impact point. Compressive wave, after reflection from the back surface, can generate tensile stresses of sufficient magnitudes to create failure near the back face.

Some investigations have focused on the response of textile fabrics without any resin matrix to high velocity impact [8, 11, 20], while others have investigated the role of resin matrix properties [21].

In examining the role of resin addition on the impact resistance of textile composite, Hsieh et al. [22] subjected Kevlar and Spectra fabrics and their fabric-reinforced composite to low velocity and high velocity impact. For both Kevlar and Spectra composites, the composites outperformed their corresponding fabric for the entire range of thicknesses tested. Examination of the specimens revealed that more fibers were broken in the composites than in the fabrics. Noting a primary role of fiber straining in energy absorption under impact, the observed fact may be used to explain the trend of superior impact performance in composites, at least in a qualitative sense.

In the case of multi-ply composite laminates reinforced with collimated Spectra webs as well as woven Spectra fabrics, the failure modes were closely observed by Lee et al., under the high velocity impact [21]. Both the angle-plied unidirectional Spectra fiber reinforced laminates and the Spectra woven fabric reinforced composites were found to exhibit sequential delamination, cut-out of a plug induced by through-the-thickness shear, and combined modes of shear cut-off and tensile failure of fibers. In the angle plied laminates, the fibers experiencing the initial impact failed in shear or by cutting, which seemed to occur at the edge of the impactor. In those cases where fiber damage was visible in the back layers of the laminate, the mode closely resembled tensile failure. However, many of the thin laminates appeared to allow the impactor to penetrate by moving fibers laterally or through fiber pull out rather than straining the fibers to break.

In contrast to the case of angle-plied unidirectional fiber reinforced composite laminates, Spectra fabric reinforced composites exhibited much less lateral movement of reinforcing fibers during the penetration of the impactor [21]. Even in thin panels, fibers apparently failed due to shear or a cutting mode in the plies close to the striking surface and in tension at the rear of a completely penetrated panel. The presence of a strip of finite width of the first lamina pushed forward by

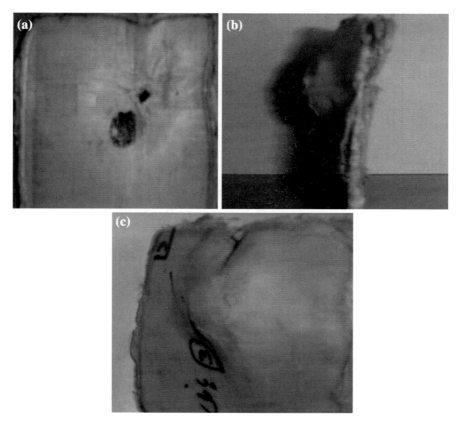

Fig. 3 Pictures of HPPE fibers cross-plies composite after impact, **a** the cavity in front of composite, **b** delamination of fibers layers and **c** back-face deformation the form of a bulge

the impactor could not be observed in the Spectra fabric reinforced composites. Instead, the delamination zones were observed preferentially along the two reinforcement directions of the woven fabric. However, these damage zones were closely integrated with a circular zone of delamination around the perforated area. The occurrence of a less anisotropic pattern of delamination was understandable considering the presence of resin-rich pockets between the reinforcing layers as well as a greater constraint to matrix crack propagation parallel to the fibers.

In our laboratory, we also investigated the high impact properties of high performance polyethylene fiber/kraton® cross-ply composites. Composites were made using a high temperature high pressure compacting process. High impact properties examined via gas-gun apparatus (see Fig. 1b) by an impactor with hemispherical tip, weight of 11 g, 9 mm diameter, and ca. 220 m/s velocity.

It is noticed that the ability to apply hot compaction to HPPE fibers without causing loss of crystallinity and orientation of the fibers and translate them into the composite enables the composite to have outstanding properties and is currently

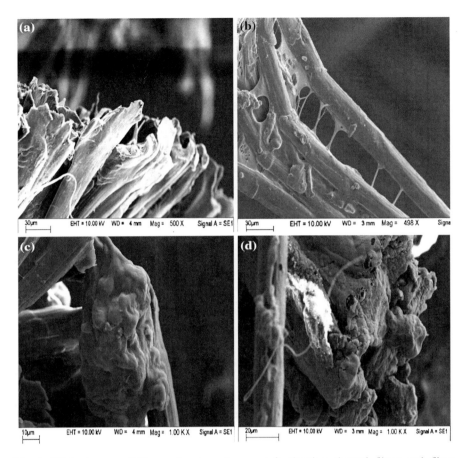

Fig. 4 SEM pictures of fibers after impact, **a** and **b** showing sheared fibers and fibers delamination accompanied by matrix, **c** and **d** melted and deformed fibers

utilized for fabrication of high strength, high modulus, and high impact resistant composites [5, 23–27]. It is well known that at elevated temperatures polyethylene is rubbery and can be subjected to considerable stretching. If layers of polyethylene fibers are properly confined to prevent shrinkage exposed to an appropriate heating and stretching sequence, consolidated and allowed to cool, it is feasible to produce a thick, rigid, and high impact resistance polyethylene fiber-reinforced composite.

After investigation of high performance polyethylene fiber cross-ply composites to high velocity impact loading, it was found that during high-velocity impact test a shock wave is created on specimen due to kinetic energy of the impactor. A part of this energy is absorbed by the fibers with shear and tensile failure mechanism and some of the energy could be absorbed by delamination of fiber layers and the rest is transmitted to the back of the specimen and forms a bulge (see Figs. 3b, c).

Figure 4a–d shows SEM pictures of fibers after impact. Failure of the fibers was via a shear mechanism, which is indicative of a high energy impact.

6 Conclusion

Polymer fiber-reinforced composites have received great attention for the past three decades in engineering applications, especially where light-weight structures are required.

It could be concluded from the experimental results and literature review that there are many parameters that affect the initiation and growth of impact damage in fiber-reinforced composites. In general, for low velocity impacts, damage starts with the creation of a matrix shear or tensile cracks and matrix cracks induce delamination at interfaces between adjacent plies and initiate a pattern of damage evolution either from bottom up or from top down. In high velocity impacts energy could be absorbed by shear and tensile failure, matrix cracking, fibers delamination, and back-face deformation in the form of a bulge.

Acknowledgments The authors would like to acknowledge gratefully the financial support of the Young Researchers Club (YRC) of Islamic Azad University, South Tehran Branch.

References

1. Mallick, P.K.: Fiber-Reinforced Composites: Materials Manufacturing and Design. Marcel Dekker, New York (1998)
2. Peters, S.T.: Handbook of Composites, 2nd edn. Chapman & Hall, London (1998)
3. Hearl, J.W.S.: High Performance Fiber. Woodhead Publishing Limited in association with The Textile Institute, Cambridge, England (2000)
4. Kavesh, S., Prevorsek, D.C.: Ultra high strength high modulus polyethylene spectra fibers and composites. Int. J. Polym. Mater. **30**(2), 15 (1995)
5. Smith, P., Lemstra, P.J., Kalb, B., et al.: Ultra high strength polyethylene filaments by solution spinning and hot drawing. Polym. Bull. **1**(11), 733 (1979)
6. Yan, R.J., Hine, P.J., Ward, I.M., et al.: The hot compaction of Spectra gel–spun polyethylene fibre. J. Mater. Sci. **32**(18), 4821 (1997)
7. Tao, Xu, Farris, R.J.: Comparative studies of ultra high molecular weight polyethylene fiber reinforced composites. Polym. Eng. Sci. 1544 (2007)
8. Tao, Xu, Farris, R.J.: Optimizing the impact resistance of matrix-free spectra fiber-reinforced composites. J. Comp. Mater. **39**(13), 1203 (2005)
9. Mallick, P.K.: Composites Engineering Handbook. Marcel Dekker, New York (1997)
10. Smith, J.C., Blandford, J.M., Towne, K.M.: Stress-strain relationships in yarns subjected to rapid impact loading: part VIII. Shock waves, limiting breaking velocities, and critical velocities. Tex. Res. J. **32**(1), 67 (1962)
11. Cunniff, P.M.: An analysis of the system effects in woven fabric under ballistic impact. Tex. Res. J. **62**(9), 495 (1992)
12. Delfosse, D., Pageau, G., Bennett, A., et al.: Instrumented impact testing at high velocities. J. Comp. Tech. Res. **15**(1), 38 (1993)

13. Jenq, S.T., Jing, H.S., Chung, C.: Predicting the ballistic limit for plain woven glass/epoxy composite laminate. Int. J. Impact Eng. **15**(4), 451 (1994)
14. Ambur, D.R., Prasad, C.B., Waters, W.A.: A dropped weight apparatus for low speed impact testing of composite structures. Exp. Mech. **35**(1), 77 (1995)
15. Wu, E., Liau, J.: Impact of unstitched laminates bi-line loading. J. Comp. Mater. **28**(17), 1641 (1994)
16. Jih, C.J., Sun, C.T.: Prediction of delamination in composite laminates subjected to low velocity impact. J. Comp Mater **27**(7), 684 (1993)
17. Choi, H.Y., Chang, F.K.: Model for predicting damage in graphite/epoxy laminated composites resulting from low velocity point impact. J. Comp. Mater. **26**(14), 2134 (1992)
18. Abrate, S.: Impact on Composite Structures. Cambridge University Press, Cambridge, England (1998)
19. Prevorsek, D.C., Kwon, Y.D., Chin, H.B.: Analysis of the temperature rise in the projectile and extended chain polyethylene fiber composite armor during ballistic impact and penetration. Polym. Eng. Sci. **34**, 141 (1994)
20. Zhu, G., Goldsmith, W., Dharan, C.K.H.: Penetration of laminated kevlar by projectiles. I. Experimental investigation. Int. J. Solids Struct. **29**(4), 399 (1992)
21. Zhu, G., Goldsmith, W., Dharan, C.K.H.: Penetration of laminated kevlar by projectiles. II. Analytical model. Int. J. Solids Struct. **29**(4), 421 (1992)
22. Roylance, D., Wilde, A., Tocci, G.: ballistic impact of textile structures. Text. Res. J. **43**, 34 (1973)
23. Lee, B.L., Song, J.W., Ward, J.E.: Failure of spectra polyethylene fiber-reinforced composites under ballistic impact loading. J. Comp Mater **28**(13), 1202 (1994)
24. Hsieh, C.Y., Mount, A., Jang, B.Z., et al.: Response of polymer composites to high and low velocity impact. 22nd International SAMPE technical conference, Boston, 6–8 November
25. Saligheh, O., Eslami Farsani, R., Khajavi, R.: The effect of post hot compaction on crystallinity and thermal behavior of ultra-high molecular weight polyethylene fiber laminates. J. Macromol. Sci. Part B: Phys. **48**(4), 766 (2009)
26. Ajji, A., Ait-Kadi, A., Rochette, A.: Polyethylene-ultra high modulus polyethylene short fibers composites. J. Comp. Mater. **26**, 121 (1992)
27. Olley, R.H., Bassett, D.C., Hine, P.J., et al.: The hot compaction of high modulus melt-spun polyethylene fibres. J. Mater. Sci. **28**, 110 (1993)

Dynamic Fracture Toughness of Composite Materials

C. Rubio-González, J. Wang, J. Martinez and H. Kaur

Abstract The Dynamic fracture toughness K_{Id}, is determined for unidirectional carbon-epoxy and glass-epoxy composite materials, by means of an experimental–numerical method. An instrumented Hopkinson bar is used to make the tests with pre-cracked specimens loaded on a three point bending configuration. Specimen receives a sudden impact load that generates the opening of the crack faces. Dynamic pulses registered on the incident and transmitted bars are used to determine the load history applied on the specimen. A strain gage is placed on the specimen to register the wave propagation and therefore to determine the onset of the crack growth. This load history is then used in a numerical analysis done by the ABAQUS software to determine the Dynamic Stress Intensity Factor time evolution. Knowing the time to fracture it is possible to estimate the Dynamic Fracture Toughness K_{Id}. For the composite material specimens, the tests were made for different impact velocities.

1 Introduction

Composite materials are generally used because they have desirable properties such as high strength per mass ratio, which could not be achieved by either of the constituent materials acting alone. They have unique advantages over monolithic

C. Rubio-González (✉) · J. Martinez
Centro de Ingeniería y Desarrollo Industrial, Pie de la Cuesta 702,
Desarrollo San Pablo, 76130 Querétaro, QRO, Mexico
e-mail: crubio@cidesi.mx

J. Wang
Department of Engineering Technology and Industrial Distribution,
Texas A&M University, College Station, TX, USA

H. Kaur
Department of Mechanical Engineering, A&M University,
College Station, TX 77843-3367 USA

materials, such as high strength, high stiffness, long fatigue life, low density, and adaptability to the intended function of the structure. Additional advantages include: corrosion resistance, wear resistance, appearance, temperature-dependent behavior, thermal stability, thermal insulation, thermal conductivity, and acoustic insulation. Despite their importance, dynamic fracture mechanics studies and experimental data for understanding the mode of failure and energy absorption mechanism of laminated composite plates are very limited. Static fracture behavior of composites has been extensively studied by many authors. By contrast, dynamic fracture toughness in composites has not received as much attention.

The evaluation of the dynamic fracture toughness of engineering materials is important for the assurance of the integrity and safety of structural components subjected to impact loading. Up to now, there has been no completely adequate method to characterize and measure dynamic fracture toughness of solid materials owing to both the difficulties in dynamic fracture theory and experimental techniques. There are discrepancies from the available reports in the rate effect on initiation fracture toughness and dynamic delamination fracture toughness of polymer composites [1]. At least three areas are open for the dynamic response of composite materials: higher delamination crack speeds, dynamic delamination initiation fracture, and the relation between initiation fracture toughness and dynamic toughness at high speed of crack propagation [1].

Dynamic fracture and delamination of unidirectional graphite/epoxy composites has been investigated for end-notched flexure (ENF) and center-notched flexure (CNF) pure mode II loading configurations using a modified split Hopkinson pressure bar in [2]. Static and dynamic fracture of interfaces between orthotropic (unidirectional glass fiber reinforced epoxy composite) and isotropic materials were studied using photoelasticity in [3]. A method to simultaneously measure fracture initiation toughness, fracture energy and fracture propagation toughness for mode-I fractures in split Hopkinson pressure bar (SHPB) with a notched semi-circular bend specimen has been proposed in [4]. Another procedure for measuring the dynamic fracture-initiation toughness of materials, based on three-point bending tests at high loading rates has been used in [5]. The experimental device is a modification of the classical split Hopkinson pressure bar. An instrumented Charpy apparatus has also been used for measuring the dynamic fracture toughness of materials [6]. A modified end-notched flexure (ENF) specimen was used to determine Mode-II-dominated dynamic delamination fracture toughness of fiber composites at high crack propagation speeds in [7]. The results showed that the dynamic fracture toughness is basically equal to the static fracture toughness and is not significantly affected by crack speeds up to 1,100 m/s [7]. Yokoyama [8] has proposed a novel impact test procedure for determining the dynamic fracture initiation toughness at high loading rates. The method uses a special arrangement of the split Hopkinson bar to measure the dynamic loads applied to pre-cracked bend specimens.

The aim of this paper is to investigate the dynamic fracture toughness K_{Id} at different loading rates of specimens made of carbon-epoxy and glass-epoxy composite materials. Only unidirectional materials were considered on the study. To this end the method proposed in [8], originally applied on isotropic materials; has been extended and employed to test composite materials. The impact bend test apparatus is based on the split Hopkinson pressure bar to measure dynamic loads acting on a precracked bend specimen. The dynamic stress intensity factor history for the bend specimen is evaluated by means of a dynamic finite element analysis based on the measured dynamic loads. The time of crack initiation is measured using a small strain gage mounted on the side of the specimen near the crack tip. The value of K_{Id} is determined from the critical dynamic stress intensity factor at the onset of crack initiation. The K_{Id} values obtained for two composite materials are compared with the corresponding values determined under quasi-static loading conditions.

2 Experimental Procedure

2.1 Test Materials and Specimen Preparation

Two composite materials were tested: carbon-epoxy and glass-epoxy. Composite laminates were obtained by curing DA 4518U Unidirectional Carbon Epoxy Prepreg (41 layers) and DA 409U/S2-Glass Unidirectional S2-glass Epoxy Prepreg (23 layers); both supplied by the company APCM.

A series of tension tests were performed to determine the material properties of the carbon-epoxy and glass-epoxy composite materials used on the experiments. The material was assumed to be transversely isotropic. The following properties were measured: Young's modulus E_1 and E_2, Poisson ration v_{12} and shear modulus G_{12}; 1-direction is along the fibers and 2-direction is perpendicular to fibers. The tension test was conducted according to recommendations of ASTM D3039-08 [9] and performed on a MTS810 testing machine. Tests were performed on several specimens at different fiber orientation; where θ is the angle between the fibers and the direction of the load, see Fig. 1. With $\theta = 0°$, E_1 is found; with $\theta = 90°$, E_2 is measured, and G_{12} is found using off-axis tests. Although the off-axis test is not a standard it is a useful method for G_{12} determination [10]. A summary of the mechanical properties including density at room temperature is presented in Table 1. Constituent content of both composite materials were determined according with ASTM D3171 [11] and the results are shown in Table 2.

Figure 2 shows the specimen geometry and dimensions. Figure 2b is a photograph of a specimen used in dynamic tests which shows the location of the strain gage near the notch tip. A thin slit of length 10 mm was cut on the specimens by a diamond disk to simulate the crack (disk dimensions were, diameter 127 mm and thickness 0.38 mm). The relative dimensions of bend specimens (width $W = 20$ mm and length $L = 100$ mm) were determined in accordance with the ASTM E399 [12]. Specimen thickness was $B = 6.3$ mm for carbon-epoxy and $B = 4.8$ mm for glass-epoxy laminates. A fixed support midspan $S = 80$ mm ($S/W = 4$) was used through this investigation.

Fig. 1 Stress–strain curves of fiber reinforced composite materials under quasi-static tensile loading conditions and different fiber orientation with respect to loading direction. **a** Carbon-epoxy, **b** Glass-epoxy

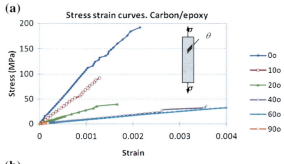

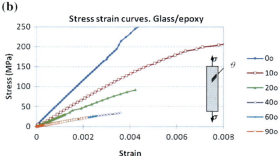

Table 1 Mechanical properties of the composite materials used on the experiments

Property	Carbon/epoxy	Glass/epoxy
E_1 (GPa)	107.7	56.7
E_2 (GPa)	8.1	10.6
v_{12}	0.34	0.36
G_{12} (GPa)	3.85	3.55
ρ (Kg/m^3)	1505.8	1795.62

Table 2 Constituent content of the composite materials used on the experiments

	Carbon-epoxy	Glass-epoxy
Fiber content (%v)	52.6	55.1
Matrix content (%v)	44.4	42.7
Void content (%v)	2.7	2.2

2.2 Quasi-Static Fracture Tests

Standard K_{IC} tests were conducted on bend specimens (Fig. 2) in a MTS810 testing machine. A notch with length of 10 mm was made by using a diamond disk. The load versus load-line displacement was recorded. Fracture toughness K_{IC} was calculated from the standard formula for the three-point bend specimen [12] as

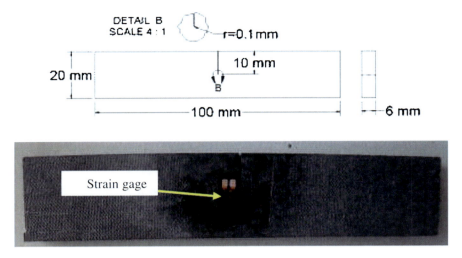

Fig. 2 Bend specimens used for fracture toughness experiments. Dimensions in mm. Specimen thickness was 6.3 mm for carbon-epoxy and 4.8 mm for glass-epoxy

$$K_Q = \frac{P_Q S}{BW^{3/2}} f\left(\frac{a}{W}\right) \quad (1)$$

$$f\left(\frac{a}{W}\right) = \frac{3\sqrt{\frac{a}{W}}}{2\left(1 + 2\frac{a}{W}\right)\left(1 - \frac{a}{W}\right)^{3/2}} \left[\left(1.99 - \frac{a}{W}\right)\left(2.15 - 3.93\left(\frac{a}{W}\right) + 2.7\left(\frac{a}{W}\right)^2\right)\right] \quad (2)$$

where K_Q is the tentative fracture toughness and P_Q is the applied load at fracture initiation.

2.3 Impact Bend Test Apparatus and Dynamic Fracture Tests

Figure 3 shows a schematic illustration of the impact bend test apparatus used for the dynamic fracture experiments. The apparatus consists mainly of an air gun, a projectile (striker bar), three Hopkinson pressure bars (one incident and two transmitters), a velocity measuring device and recording equipment. The pressure bars (1,800 mm long) and the striker bar (152.4 mm long) were made of 19 mm diameter high strength steel (Maraging C300). The bend specimen was placed between the incident and two transmitter bars. The specimen was rapidly loaded to fracture by means of a concentrated transverse load applied at midspan on the uncracked surface of the specimen. The split Hopkinson bar technique was employed to measure the dynamic load exerted on the specimen. The impact of the striker bar into the face of the incident bar develops the longitudinal compressive pulse $\varepsilon_i(t)$ that is propagated along this bar. Part of the compressive incident pulse $\varepsilon_i(t)$ is transmitted through the specimen, causing its fracture and into the transmitted bars as a compressive pulse $\varepsilon_t(t)$, while part of which is reflected back to the incident bar as a tensile pulse $\varepsilon_r(t)$. The incident load $P_i(t)$ and the transmitted load $P_t(t)$ are then determined from the one-dimensional theory of elastic wave propagation as [13]

$$P_i = EA[\varepsilon_t(t) + \varepsilon_r(t)] \tag{3}$$

and

$$P_t = EA\,\varepsilon_t(t) \tag{4}$$

where E and A are, respectively, Young's modulus and the cross sectional area of the pressure bars. In the foregoing derivations, the recorded strain pulses are assumed shifted along the time axis so as to be coincident at the specimen surface. The time is taken to be zero at the instant when the incident wave front arrives at the uncracked surface of the specimen. In the experiments, the three strain pulses were monitored by means of three strain gages having a gage length of 1.6 mm (VISHAY CEA 06 062 UW 120) bonded in the incident and transmitter bars using M Bond 200 adhesive. The first strain gage was located 900 mm from the impact end of the incident bar; and the other two strain gages were placed 900 mm from the ends of the respective transmitter bars, see Fig. 3. The strain gage placed on the specimen close to the notch tip was VISHAY CEA 13 032 UW 120. The output signals of the strain gages were conditioned by conditioners VISHAY 2311 and then visualized and stored into an oscilloscope Agilent model Infinium 5483 where the signals were recorded at a sampling rate of 1 µs.

The rounded ends of the incident and transmitter bars had no significant effect on the wave propagation in the respective pressure bars. The dynamic loads P_i and P_t calculated from Eqs. 4 and 5 were used for the accurate evaluation of the dynamic stress intensity factor of bend specimens through a finite element analysis. Dynamic fracture toughness K_{Id} was determined from the critical dynamic stress intensity factor at the onset of crack initiation. The bend specimens used in K_{IC} test (quasi-static test) have the same geometry as those used in the dynamic fracture tests (Fig. 2).

3 Results and Discussion

To determine quasi-static fracture toughness, K_{IC}, the load versus load line displacement curve was registered for each test. A typical curve for carbon-epoxy and glass-epoxy is shown in Fig. 4a and b, respectively. Tests were performed under load control. According to the ASTM E399 standard [12], a P_Q load value is determined from the intersection of the load—load line displacement curve with a straight line of 95% of the slope in the linear region. Such intersection is shown in

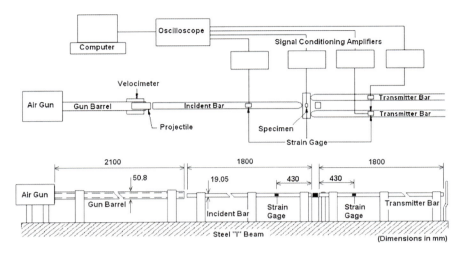

Fig. 3 Impact bend test apparatus employed for dynamic fracture experiments and block diagram of the recording system

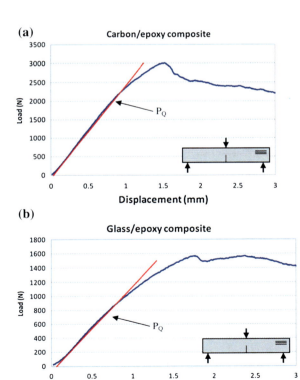

Fig. 4 Typical load–displacement curve to determine P_Q and the quasi-static fracture toughness. **a** Carbon-epoxy, **b** Glass-epoxy

Fig. 4. Values of P_Q are used in Eq. 2 to calculate K_{IC} giving the following results. For carbon-epoxy: 26.9 MPa(m)$^{1/2}$ and for glass-epoxy: 13.9 MPa(m)$^{1/2}$. In both cases fibers were perpendicular to the crack as illustrated in Fig. 4. It is worth noting that for the carbon-epoxy composite the value of P_Q represents approximately 75% of the maximum load reached during the test, while for glass-epoxy that value represents 55%. A more nonlinear behavior is observed in the glass-epoxy composite.

Figure 5 shows typical oscilloscope records from the dynamic fracture test on a carbon-epoxy composite specimen. The incident pulse has a duration time of about 66 μs. Figure 6 shows the same information as that provided by Fig. 7 but in the appropriate scale. Figure 7 presents a typical incident load ($P_i(t)$) obtained from Eq. 4 for both composite materials. The signal of the small strain gage attached near the crack tip is shown also in Fig. 5. The location of the strain gage was determined so that the dynamic strain could be measured closer to the crack tip.

Note that the incident pulse amplitude is almost the same on each material in Fig. 6; this was expected since the amplitude depends only on the projectile velocity, according with the following equation [13]

$$P = \sigma A = \frac{1}{2} \rho c v A \tag{5}$$

where ρ is the mass density, c is the dilatational wave speed and A is the cross section area, these parameter are associated with the incident pressure bar; and v is the projectile velocity.

The finite element technique was applied to evaluate the dynamic stress intensity factor $K_I(t)$ for the specimen using the commercial software ABAQUS. Because of symmetry, only half of the specimen was modeled with 8-node isoparametric quadrilateral elements, as shown in Fig. 8, singular elements were used around the crack tip. It was assumed in the finite element analysis that: (a) the crack tip is under the conditions of plane stress; and (b) the pre-crack is stationary during the dynamic fracture test. Since the transmitted pulses occurred at about 60 μs and the time to fracture is less than that time, the applied load to the finite element model was only the incident load $P(t)$ given by Eq. 4, that is, the transmitted load $P_t(t)$ had no effect; thus, the model was unsupported during the period of analysis. The dynamic load given in Fig. 7 was used as the input (or time dependent) boundary conditions (the load applied was $P_i(t)/2$ because only half of the specimen is modeled). The elastodynamic response of the pre-cracked specimen was obtained by direct integration of the equations of motion.

The dynamic stress intensity factor $K_I(t)$ was computed by simply comparing the dynamic crack opening displacement $v(t)$ of a node close to the crack tip with its static value according to the following equation [14],

$$K_I(t) = \frac{\pi}{\sqrt{2\pi r}} v(t) \frac{\omega_1 \omega_2}{\omega_1 + \omega_2} \frac{1}{a_{22}} \tag{6}$$

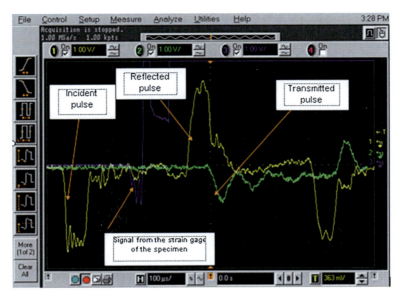

Fig. 5 Typical oscilloscope measurements from dynamic fracture tests of carbon-epoxy composite specimen (projectile impact velocity $v = 15.7$ m/s)

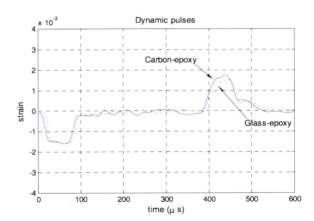

Fig. 6 Typical pulses arranged in the appropriate scale. Projectile impact velocity $v = 15.7$ m/s for carbon-epoxy and $v = 15.6$ m/s for glass-epoxy

where ω is related with μ according to

$$\mu_j = i\omega_j \quad j = 1, 2 \qquad (7)$$

and μ_1 and μ_2 are roots of

$$a_{11}\mu^4 + (2a_{12} + a_{66})\mu^2 + a_{22} = 0 \qquad (8)$$

with a_{ij} being linked with the elastic constants according to

$$a_{11} = 1/E_1 \quad a_{12} = -v_{12}/E_1 \quad a_{66} = 1/G_{12} \quad a_{22} = 1/E_2 \qquad (9)$$

Fig. 7 Dynamic load $P_i(t)$ applied to the bend specimen according to Eq. 4

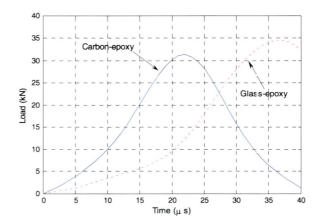

Fig. 8 Finite element mesh used to determine dynamic stress intensity factor $K_I(t)$ on bend specimens. Only one half is modeled and loaded with $P_i(t)/2$

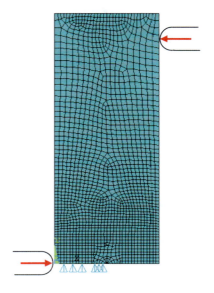

for plane stress; r is the distance from the crack tip to the node where transverse displacement $v(t)$ is determined.

Figure 9 displays the time variation of the nodal transverse displacement $v(t)$ (perpendicular to the crack faces) of the unsupported end of the impacted specimen for both materials. A closer view of the time when nodal displacements grows up is also shown in Fig. 9 for carbon-epoxy. That time is approximately 4.8 µs which coincides with the arrival of the dilatational wave at the node under analysis (located at 11 mm from the uncracked edge), this is verified calculating

Fig. 9 Time evolution of transverse displacement after the specimen impact. The displacement component $v(t)$ corresponds to a node located on the crack face at 1 mm from the crack tip. This displacement is used to determine $K_I(t)$ according with Eq. 6

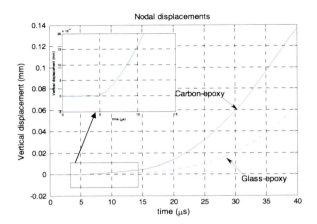

Table 3 Wave velocities. Dilatational C_d, shear C_s, and Rayleigh C_R

	Carbon/epoxy		Glass/epoxy	
	Propagation parallel to the fibers	Propagation perpendicular to the fibers	Propagation parallel to the fibers	Propagation perpendicular to the fibers
C_d (m/s)	8494.15	2329.00	5688.66	2459.36
C_s (m/s)	1588.57	1598.99	1406.07	1406.07
C_R (m/s)	1575.34	1574.92	1390.92	1385.11

the different waves velocities [15] for the composites used in this work. The velocities are shown in Table 3. Nodal displacement is zero until that time and this effect is well captured by the numerical solution. Figure 10 represents the $K_I(t)$ curve for the unsupported specimen calculated with Eq. 6 and using the displacement history $v(t)$ given in Fig. 9. It is important to note that during the early stages of the dynamic fracture test, $K_I(t)$ is unaffected by the support condition of the specimen. The value of the dynamic fracture toughness K_{Id} was determined from the critical dynamic stress intensity factor at the instant of crack initiation. The time of crack initiation, t_c, was determined from the signal output from the strain gage attached to the specimen, for this case it was about 29 μs giving approximately $K_{Id} = 38.4$ MPa(m)$^{1/2}$ for carbon-epoxy composite, and 39 μs giving approximately $K_{Id} = 29.3$ MPa(m)$^{1/2}$ for glass-epoxy composite as shown in Fig. 10.

Figure 11a, b shows the dynamic fracture toughness of carbon-epoxy and glass-epoxy, respectively for some impact projectile velocities. Quasi-static fracture toughness is also shown for comparison. Dynamic fracture toughness values for both materials are greater than those for quasi-static toughness. Also, no significant variation of dynamic fracture is observed when changing the projectile velocity (associated with strain rate), at least for the ranges used on the experiments.

Fig. 10 Stress intensity factor time evolution $K_I(t)$ for impacted bend specimens. **a** Carbon-epoxy, **b** Glass-epoxy

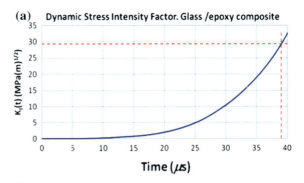

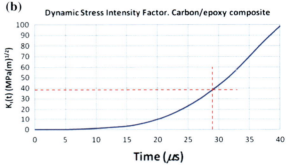

Fig. 11 Dynamic fracture toughness of composite materials at different projectile velocities, quasi-static toughness is also shown for comparison. **a** Carbon-epoxy, **b** Glass-epoxy

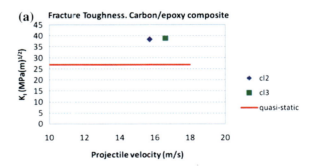

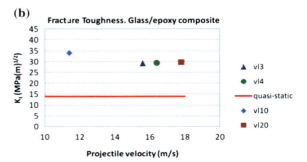

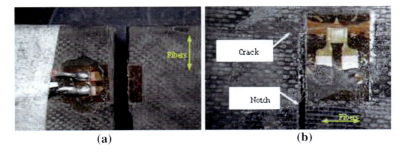

Fig. 12 Fracture of carbon epoxy composite specimens. **a** Fibers parallel to the notch, **b** fibers perpendicular to the notch

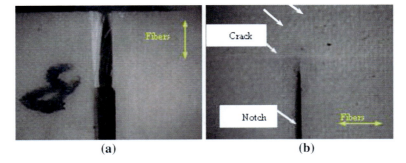

Fig. 13 Fracture of glass epoxy composite specimens. **a** Fibers parallel to the notch, **b** fibers perpendicular to the notch

Figures 12 and 13 show specimens after dynamic fracture tests. Note that when fibers are parallel to the notch, the specimens split completely. When the fibers are perpendicular to the notch, a crack is generated on the notch tip and it propagates along the fiber direction, as shown in Figs. 12b and 13b. Photographs of cracked specimens with fibers parallel to the notch are shown just to compare the failure mode with that exhibited by specimens with fibers perpendicular to the notch, although no further analysis for specimens with fibers perpendicular to the notch is presented in this work.

4 Conclusions

Dynamic fracture toughness has been evaluated for unidirectional carbon-epoxy and glass-epoxy composite materials with fibers perpendicular to the notch. A special set-up based on the split Hopkinson bar apparatus has been used to measure dynamic loads acting on notched bend specimens. The dynamic stress intensity factor history for the specimens was determined using the finite element method based on the measured dynamic loads. Since the time of crack initiation was recorded, then, the dynamic fracture toughness of the composite samples was estimated. Dynamic fracture toughness is greater than quasi-static toughness for both materials. It was observed that, except for some experimental scatter, dynamic fracture toughness K_{Id} of glass epoxy is not affected by projectile velocity (strain rate). For carbon-epoxy the results are not enough to make a conclusion regarding the effect of strain rate in K_{Id}.

Acknowledgments This research was supported by the Texas A&M University-CONACYT Collaborative Research Grant Program.

References

1. Sun, C.T., Hang, C.: A method for testing interlaminar dynamic fracture toughness of polymeric composites. Compos. B **35**, 647–655 (2004)
2. Nwosu, S.N., Hui, D., Dutta, P.K.: Dynamic mode II delamination fracture of unidirectional graphite/epoxy composites. Compos. B **34**:303–316 (2003)
3. Shukla, A., Chalivendra, V.B., Parameswaran, V., Lee, K.H.: Photoelastic investigation of interfacial fracture between orthotropic and isotropic materials. Opt. Lasers. Eng. **40**(4), 307–324 (2003)
4. Chen, R., Xia, K., Dai, F., Lu, F., Luo, S.N.: Determination of dynamic fracture parameters using a semi-circular bend technique in split Hopkinson pressure bar testing. Eng. Fract. Mech. **76**, 1268–1276 (2009)
5. Rubio, L.: Determination of dynamic fracture-initiation toughness using three-point bending tests in a modified Hopkinson pressure bar Exp. Mech. **43**(4), 379–386 (2003)
6. Fengchun, J., Ruitang, L., Xiaoxin, Z., Vecchio, K.S., Rohatgi, A.: Evaluation of dynamic fracture toughness Kid by Hopkinson pressure bar loaded instrumented Charpy impact test. Eng. Fract. Mech. **71**, 279–287 (2004)
7. Tsai, J.L., Guo, C., Sun, C.T.: Dynamic delamination fracture toughness in unidirectional polymeric composites. Compos. Sci. Technol. **61**, 87–94 (2001)
8. Yokoyama, T.: Determination of dynamic fracture-initiation toughness using a novel impact bend test procedure. Trans. ASME J. Press. Vessel Technol. **114**, 389–397 (1993)
9. ASTM, Annual book of ASTM Standards, No. D 3039—08, Standard Test Methods for Tensile Properties of Polymer Composite Materials
10. Gibson, R.F.: Principles of Composite Material Mechanics. McGraw-Hill Inc, New York (1994)
11. ASTM, Annual book of ASTM Standards, No. D 3171—09, Standard Test Methods for Constituent Content of Composite Materials
12. ASTM 2002, Annual book of ASTM Standards, v.03.01, No. E399-02, Standard Test Method for Plane Strain Fracture Toughness of Metallic Materials
13. Graff, K.F.: Wave Motion in Elastic Solids. Dover Publications, New York (1991)
14. Williams, J.G.: Fracture mechanics of anisotropic materials. In: Friedrich, K. (ed.) Application of Fracture Mechanics to Composite Materials. Elsevier, Amsterdam (1989)
15. Nayfeh, A.H.: Wave Propagation in Layered Anisotropic Media with Applications to Composites. North-Holland Series in Applied Mathematics and Mechanics, Vol. 39. North-Holland, Amsterdam (1995)

Impact Study on Aircraft Type Laminate Composite Plate; Experimental, Failure Criteria and Element Model Review

Y. Aminanda

Abstract This paper reviews the advancement of study of composite laminate subjected to impact loading for aerospace application such as Glass fiber-epoxy and Carbon fiber-epoxy unidirectional laminate plate. The impact testing set-up and equipments are overviewed in this paper including the calculation to obtain the force, velocity and energy absorbed by the laminate during impact. The initial damage can be explained by the threshold of energy absorption while the successive damage mechanism can be described by the drop of force obtained from the history of force during impact. The corresponding evidence of damage related to force drop shows the scenario of damage mechanism; matrix cracking, delamination and fiber failure. The influence of projectile and laminate parameters to the impact behavior is reviewed as well. The failure criteria for each phase of damage mechanism are described in detail. The proposed elements model to simulate the damage are overviewed as well, which can be divided into two categories; element without interface layer and with interface layer integrating degradation of mechanical properties. The exposed failure criteria and elements model are the basis of today's development of FEA simulation for impact purpose.

1 Introduction

Composite has been used widely in the industry due to its performance in term of its strength while maintaining light weight. This performance attracted especially aeronautics industry to develop the structure replacing metallic material with the

---

Y. Aminanda (✉)
Mechanical Engineering Department, Kulliyah of Engineering,
International Islamic University, Kuala Lumpur, Malaysia
e-mail: yulfian@iium.edu.my

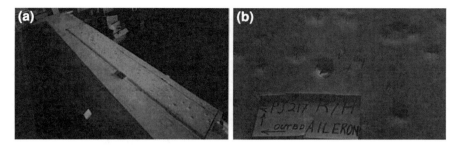

Fig. 1 Damage of aileron due to hail impact [1]

lighter one. The latest and mature composite structure utilized in this area relates to carbon fiber/epoxy matrix composite structure. Effectively, this type of structure which has been used for secondary aircraft structure is developed for maintaining high stresses acting at primary structure as in the recent aircraft such as A380 and B787. In the case of A380 aircraft, the carbon fiber/epoxy composite structure has been manufactured and designed as an integral part of wing assembly such as the whole wing box structure. For the case of B787, the whole fuselage, for example, has been fabricated integrally with composite structure without any joints.

The mechanical properties of carbon fiber/epoxy matrix were known and the standardized manufacturing process allows obtaining high strength of the material and turns this material to its maturity. However, the issue of impact loading remains the weakness of the structure, therefore the understanding and the methodology to determine the strength of composite structure subjected to impact loading always becoming the state of the art in aerospace industry.

The Aircraft faces frequently impact loading during its service. This impact loading comes from different type of events such as; bird impact, hail projection and others scenario during take-off and landing. It is also possible that the impact can be produced during maintenance visit due to tool's drop. The damage of the structure can be severed and observable as shown in Fig. 1 [1], but most of the time it is not visible. The residual strength of impacted composite structure, for example its strength in in-plane compression, could be reduced up to 50% even for the case of invisible damage. A damage tolerance damage called BVID/barely visual impact damage has been imposed in order to determine whether the impacted aircraft structure should be replaced or not to obtain a permission to flight from the authority. The ultimate goal for aircraft industry is to shorten the process of finding the damage, determination the residual strength and authorization to flight. One way in shortening the process uses finite element analysis model to simulate the behavior of composite structure subjected to impact loading. The state of the art of FEA simulation in this area varies from impact phenomena on composite plate until sub-part and part of aircraft structure (Figs. 2 and 3).

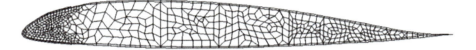

Fig. 2 FEA model for blade of helicopter (reproduction form) [2]

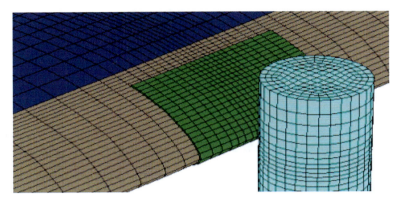

Fig. 3 FEA 3D model of rigid projectile on leading edge [2]

This paper overviews the study of impact loading on laminate composite structure that has been developed until today especially for aerospace application. The experimental set-up, observation and results analysis will be exposed including the parameters of structure, projectile and energy of impact. The proposed analytical model will be detailed.

The failure criteria in term of matrix cracking, delamination and fiber fracture will be exposed. The modeling of failure using FEA elements and including the scenario of mechanical properties degradation which are integrated to the element will be highlighted. The method of creating the elements with its failure criteria highlighted in this chapter becomes the basis of the latest element developed for commercial FEA software.

2 Experimental Study of Laminate Composite Structure Subjected to Impact Loading

In this sub-chapter the experimental set-up for impact testing is described including the determination of energy absorption, contact force during impact. The effect of composite structure, projectile, boundary conditions and energy impact will be overviewed. The mechanism of damage during impact will be described as well.

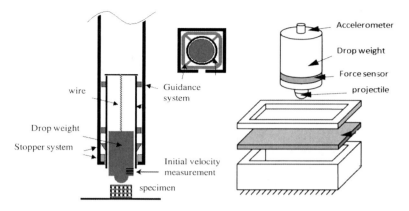

Fig. 4 Drop weight experimental set-up testing

2.1 Experimental Set-up and General Analysis of Test Results

The impact testing with low velocity of energy impact is performed using drop weight (Fig. 4) [3] or pendulum experiment set-up while for higher velocity impact gun set-up type is normally used [2]. To simulate simple case of impact loading, most of the time a spherical projectile is employed and for certain experiments cylindrical (see pyramidal) projectile is impacted on laminate plate [4, 5].

During impact, initial velocity just before impact is measured using optical or laser sensor located in the same level of specimen location. The marks at the surface of projectile drawn with a certain distance will cut the laser lines for a certain period of time which able to measure the initial velocity before impact. The initial energy of impact can be calculated then using equation below:

$$E_{impact} = \frac{m_{projectile}}{2} V_{impact}(0)^2 \qquad (1)$$

where

E_{impact}: initial impact energy
m_{impact}: Mass of projectile.
$V_{impact}(0)$: velocity of projectile just before impact

By integrating the dynamic equation, the velocity of projectile during impact can be obtained from the force data given by load cell using the equation below:

$$V_{impact}(t) = V_{impact}(0) - \frac{1}{m_{projectile}} \int_0^t F(t)dt \qquad (2)$$

where:

$V_{impact}(t)$: projectile velocity during impact
$F(t)$: Contact force during impact obtained from load cell

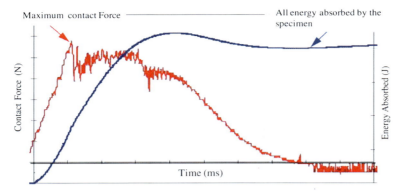

Fig. 5 Contact force and energy absorption history during impact for low impact energy

It can be noted that projectile velocity during impact can be obtained using acceleration of projectile measured by accelerometer. In practice, force data is used since it gives less noise compared to accelerometer one.

Once the projectile velocity during impact is determined, the energy absorbed by composite structure can be calculated using the following equation:

$$E_{\text{absorption}}(t) = \frac{m_{\text{impact}}}{2}(V_{\text{impact}}(0)^2 - V_{\text{impact}}(t)^2) \qquad (3)$$

Using the data available during the test, the displacement of projectile during impact can be obtained by integrating the velocity using the equation below:

$$\delta(t) = \int_0^t v(t)dt + \delta(0) \qquad (4)$$

The initial projectile displacement $\delta(0)$ can be set equal to zero corresponding to the displacement just before impact.

One example of impact test result can be found from the work that has been done in CCR-EADS [6]. The energy and contact force history during impact is shown in Fig. 5. The experimental was using hemispherical projectile of 16 mm diameter and mass of 2.76 kg. The specimen with dimension of 150 × 100 mm was clamped on its four sides in order to obtain fix type of boundary conditions. From the graph, it can be observed rebounding phenomena corresponding to force and velocity equal to zero. From the graph of energy absorption, the energy starts to be constant after contact which corresponds to the rebound phenomena.

For higher impact energy, the graph shows a different tendency as it can be observed in Fig. 6. To show the generality of result analysis, the proposed graph with higher energy was obtained from impact test result using thicker specimen such as sandwich structure. From the figure, rebounding phenomenon does not appear and energy absorbed evolution increases in function of time until the end of impact. While analyzing the graph, it can be observed that upper skin is perforated by projectile after reaching the first maximum contact force. The force becomes

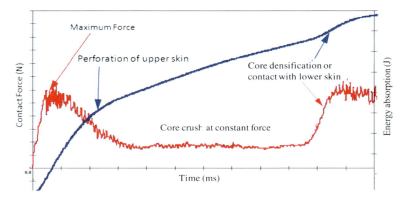

Fig. 6 Contact Force and Energy absorption history for higher energy

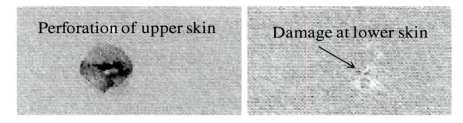

Fig. 7 Damage due to impact loading on sandwich structure with composite skins and honeycomb core

constant which correspond typically to core crush. Then the force increases again corresponding to either densification of core or contact force with lower skin of sandwich structure. From the image of damage during impact (Fig. 7), it can be shown that effectively projectile touched the lower skin without perforation. In this case the impact energy has been absorbed totally by the specimen

An interesting study has been proposed by the work of Herup [7]. The results were presented in term of ratio between initial impact energy and energy absorbed by the laminate specimen for different plies of 4, 8, 16, 32 and 48 (Fig. 8). The author stated that at least 15% of impact energy has been absorbed by the specimen even at low impact energy and without observable damage. There is a threshold characterized by sudden increment of energy absorption which relates to damage initiation. The nature of damage initiation was not specified and will be detailed in the specific sub-chapter. The threshold value increases with the thickness of laminate. After the threshold value, the ratio of energy absorption with impact energy becomes globally constant. It means if we double the impact, the energy absorption will be doubled as well. Also, the energy absorption increases with the thickness of laminate which seems normal since the dissipative energy increase with the thickness.

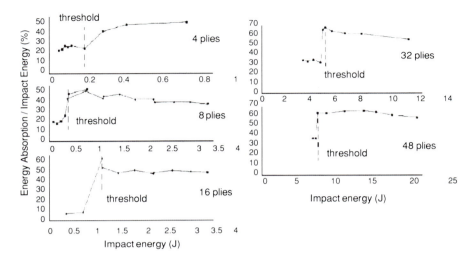

Fig. 8 Determination of initiation damage using energy approach

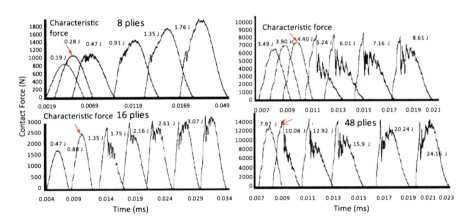

Fig. 9 Evolution of contact force for different energy impact and thickness of laminate

The initiation of damage can be interestingly studied using evolution of force during impact as shown in Fig. 9. In the figure, time is shifted voluntarily to obtain the force relatively for different level of impact energy. It can be stated that below threshold energy, the evolution of contact force increase progressively and the amplitude of contact force increases with the thickness of laminate as well. After the threshold energy, the contact force gave unsmooth curve for all thickness of laminate. The initiation of damage can be identified then from the graph and corresponding to significant decrease of contact force after reaching the maximum

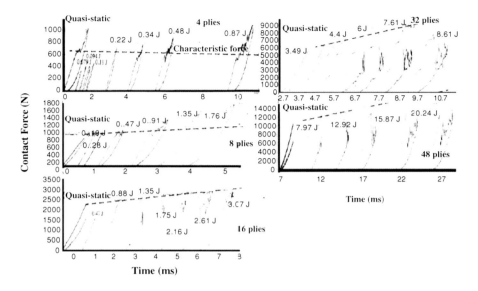

Fig. 10 Comparison of contact force between quasi-static and dynamic loading

contact force called characteristic force. It seems that the characteristic force is independent to impact energy which indicates that the initiation of damage is controlled by force generated in the area of contact.

This study shows the generality of the method for different type of specimen subjected to impact loading and the method is being used widely for industrial application.

2.2 Equivalence of Quasi-Static/Dynamic Loading

As mentioned previously, the damage happens in the contact area even for low impact energy and the invisible damage can reduce the strength of impacted composite significantly. That is the reason most of the industrial works on impact concentrates on the low energy and low velocity impact. It is interesting to study whether the low velocity impact loading can be represented by quasi-static loading where the velocity is very low and near to static loading. The quasi-static and dynamic equivalence has been performed by Herup [7]. The work compared the contact force evolution obtained from quasi-static or indentation loading with the dynamic loading one (Fig. 10). From the graph, it can be stated that independently to impact energy, the characteristic force remains in the same level for thin laminate until eight plies. The characteristic force in dynamic loading has the same level with the characteristic force of quasi-static loading. Also, the comparison of damage between quasi-static and dynamic loading as shown in Fig. 11 shows an

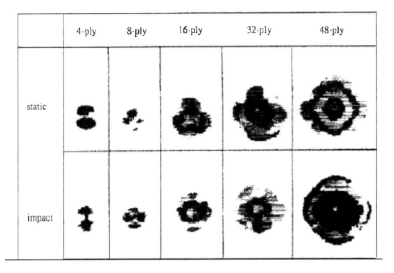

Fig. 11 Comparison of damage form and area produced by quasi-static and dynamic loading

identical form and area of damage. It can be stated that there is an equivalent quasi-static loading with low energy impact loading. The conclusion has been confirmed by the work that has been done in CCR-EADS [6]. The observation is not valid for thicker laminate due to; seemingly, the dynamic effect produced by local masse in the area of impact becomes significant to be compared by static loading.

It seems that quasi-static and dynamic loading equivalence can be applied for certain range of structure and impact energy even if its limitation is not stated clearly in my knowledge. But it is clear that for high velocity of impact, the equivalence is not valid even if the masse of impact remains light. It is appropriate then to state that there is an equivalence of quasi-static with low energy and low velocity impact loading.

2.3 Damage Mechanism

The study of damage on laminate composite structure shows that generally the damage involves three types of physics phenomena [8] which are: matrix cracking, delamination and fiber fracture as shown in Fig. 12.

The matrix cracking was due to big difference mechanical properties between matrix and fibers. It happens generally parallel with fiber direction of plies [9]. The two principal modes of cracking are generated by tension and shear stresses as shown in Fig. 13. During impact, it is observed the cracking with 45^0 direction in the area of contact until mid-height of plate due to shear stress. In the opposite ply

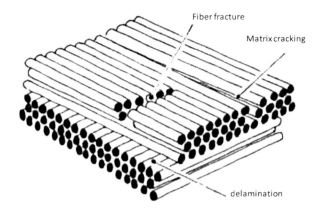

Fig. 12 Damage mechanism of laminate composite structure subjected to impact loading

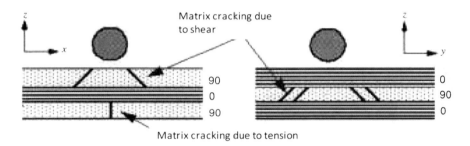

Fig. 13 Matrix cracking scenario

under contact zone, the cracking form vertical line due to normal stress produced by bending stress during impact. The cracking is frequently a combination of these two modes and also due to complex state of stress. That is the reason, the cracking criteria is associated to the coupling of different solicitation of normal and shear stress inside the ply.

According to work of Liu [10], delamination is produced from different bending rigidity between adjacent plies (Fig. 14). It can be produced only with a presence of matrix cracking, more precisely if the matrix in the adjacent plies is cracked and the interior ply is saturated by matrix cracking. The impact energy is then transmitted and propagated to the interface between plies which creates delamination [11].

Globally in term of damage mechanism, the fiber fracture is appeared only after matrix cracking and delamination. The fiber rupture can be detected on the contact area due to high magnitude of local stress and on the non-contact surface due to high bending stress [9–12].

The zone of delamination principally produced in the interface of $\pm 45^0$ of plies in elliptical or butterfly form [13] and Fig. 15 especially for laminate using fabric

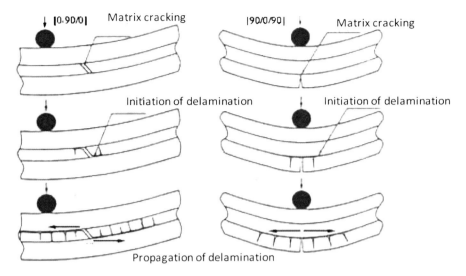

Fig. 14 Delamination due to impact loading

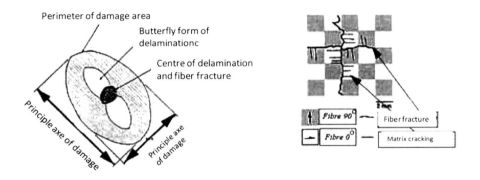

Fig. 15 Damage zone due to impact loading

ply. Tsang [14] has completed the study and indicated precisely the location of damaged fiber and matrix cracking as shown in Fig. 15. It is observed that the fiber fracture happened along the fabric direction due to significant fibers curvature at this location.

The evolution of contact force taken form impact experiment shows the damage mechanism scenario as shown in Fig. 16 [15]. The experiments have been done on carbon/epoxy fabric laminate of [0/0/0/0/0] subjected to spherical projectile with two different levels of energy 0.6 and 1 J. The specimen has been fixed by circular clamp support with diameter of 40 mm. From the graph of experimental result using low impact energy (0.6 J), damage produced by the impact is in the form of delamination as shown in the photo of damage. The contact force related to the

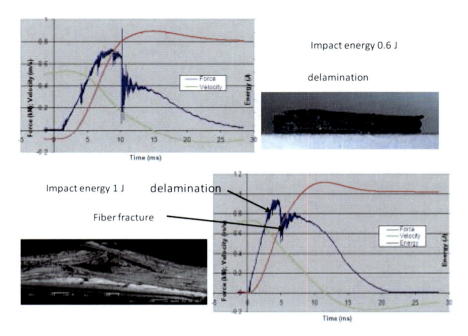

Fig. 16 Example of impact test on carbon/epoxy laminate

delamination is found around 0.7 kN. At higher impact energy of 1 J, the fiber fracture at lower surface has been observed at contact force of 0.95 kN which consequently started by delamination at 0.7 kN. After drop of contact force due to delamination, the loading is transmitted to the fibers which increase the force until the maximum before the fibers break at its turn. However, the matrix cracking which should happen before delamination is very difficult to observe and does not change the smoothness of contact force curve.

2.4 Influence of Impact Parameters

2.4.1 Projectile Form

An experimental study of impact using different type of projectile has been proposed by Mitrevski et al. [16]. Different shapes of projectile; hemispherical, ogival and conical with the same diameter of 12 mm (Fig. 17) has been used for impact testing with initial energy of 4 and 6 J on woven carbon/epoxy laminates [(45/0/45/0) and (0/45/0/45)s]. It is found that the highest energy absorbed by specimen is for conical projectile which also produce largest indentation depth. Bigger damage area seems giving highest energy absorbed for the same laminate. The maximum contact

Fig. 17 Three different shapes of projectile; hemispherical, ogival and conical

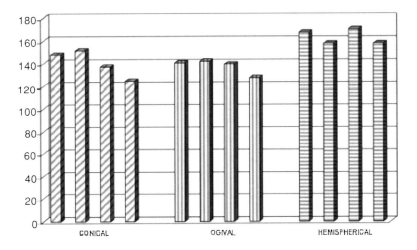

Fig. 18 Damage area due to impact using different projectile shape

force produces by hemispherical projectile with shortest duration of contact. The damage threshold load which correspond to large drop of contact force and indicates the initiation of damage is higher for hemispherical impactor followed by ogival and conical projectile. Also the damage area gives different dimension for different three projectiles as shown in Fig. 18 where hemispherical projectile produced bigger damage area followed by conical and ogival projectile.

2.4.2 Thickness and Fiber Directions

The work of Foussa [17] shows that the damage area varies linearly with damage resistance parameter. The parameter relates to a measure of the damage radius at certain angle from centre of impact. It is based on different bending strain between two adjacent plies. The damage resistance parameter is depending on the stacking sequence and damage area. The impact tests were performed using spherical projectile of 15.9 mm and 24 plies of laminate plate and the obtained damage area in function of stacking sequence is shown in Fig. 19.

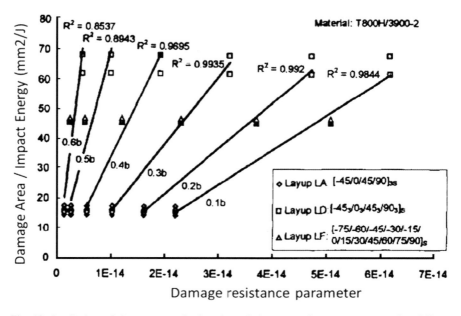

Fig. 19 Prediction of damage area in function of damage resistance parameter for different stacking sequence

It is found that the relation is highly linear, therefore the damage area can be predicted based on damage resistance parameter which is calculated based on the stacking of laminate.

The effect of thickness of laminate subjected to impact loading has been studied by Belingardi [18] using laminate of (0/60/−60) and (0/90). The number of ply for each laminate is varying from 4, 8 and 16 layers describing the different thickness of laminate. The plate is impacted using different level of energy impact; 5, 13, 38 J to obtain three different phenomena in impact such as impact with rebound, stop and perforation. It is found out that the threshold force and maximum contact force vary linearly with the thickness as shown in Fig. 20 for both stacking (0/60/−60) and (0/90). Linear relationship between force and thickness of plate seems valid only at the range of impact loading used for the experiment. Therefore, the threshold and maximum force can be predicted for a certain thickness of laminate.

3 Analytical Study of Laminate Composite Structure Subjected to Impact Loading

The first mathematical model dealt with contact force law relates to the contact force with the indentation depth. A comprehensive study on contact force for orthotropic material, quadratic contact surface and in elastic linear domain has

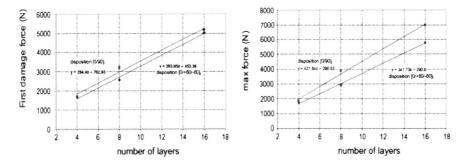

Fig. 20 Threshold and maximum force variation in function of thickness/number of layer of laminate plate

been proposed by Hertz. One of the proposed contact laws is called Meyer's Law [19] that can be expressed in the form:

$$F = k\alpha^n \tag{5}$$

where F: contact force and α: indentation depth.

The equation allows generalizing the contact force of Hetrz beyond the initial study with the value of coefficients k and n is determined adequately. For example, for laminate plate, Yang [20] proposed to use the value of k as the following:

$$k = 0.75\left(E\sqrt{R}\right) \tag{6}$$

with $\frac{1}{R} = \frac{1}{R_1} + \frac{1}{R_2}$ and $\frac{1}{E} = \frac{1-\vartheta_1^2}{E_1} + \frac{1}{E_2}$.

R_1, E_1, and v_1 are the radius, Young's modulus and Poisson's ratio of projectile while R_2 and E_2 are the damage radius and transverse stiffness of the structure. Other work proposed the equation as the following:

$$F = k\alpha^{1.5n} \text{ at loading phase} \tag{7}$$

$$F = F_m\left(\frac{\alpha - \alpha_0}{\alpha_m - \alpha_0}\right)^q \text{ at unloading phase} \tag{8}$$

Where subscription m indicates the maximum value and α_0 indicates residual of indentation which equals to zero below certain threshold value of $\alpha_{critical}$. The value of $\alpha_{critical}$ equals to 0.1016 mm for carbon/epoxy laminate plate. An indentation test on laminate plate allows determining the value of $n = 1.5$ and $q = 2.5$ which match in general with the coefficients for solid contact. The equation is valid also for general case of indentation such as for sandwich structure with carbon/epoxy $(0_2/90_2/0_2)$ skin with certain value of n, $\alpha_{critical}$ and q.

Recent work by Choi [21] developed the contact law by proposing linearized contact law with the equation below:

$$F = k_1 \alpha \tag{9}$$

with $k_1 = F_m^{\frac{1}{3}} k^{\frac{2}{3}}$.

K_l means for contact coefficient of the linearized contact law and F_m for the predicted maximum contact force. The difference with the previous law is that the value of F_m needs to be predicted first. It can be predicted using FEA model for example. The method, since it is linear, becomes very robust and fast in term of computation time. The result has been validated for 10×10 cm of laminate plate $(90/45/90/45/90)_{2s}$ subjected to indentation with different boundary conditions at its four sides.

The contact law model proposed above did not take into account in the equation the stacking sequence and anisotropic or orthotropic behavior and is valid only for small displacement. To overcome the limitations, Aboussaleh [22] proposed that the contact force can be represented as double Fourier series. A function of Green was constructed to solve the governing equation of contact force. With the hypothesis that the tangential stresses vary in the thickness of plate following a specific law and in absence of load at the external surface, the equations below can be drawn.

$$\tau_{xz} = \left(\alpha_z^1 f_1(z) + \beta_z^1\right) \emptyset(x, y) \tag{10}$$

$$\tau_{yz} = \left(\alpha_z^2 f_2(z) + \beta_z^2\right) \varphi(x, y) \tag{11}$$

Where $\emptyset(x, y)$ and $\varphi(x, y)$ are the arbitrary functions, $f_1(z)$ and $f_2(z)$ are the function characterizing the variation law of tangential stresses with thickness of laminate. β_z^1 and β_z^2 are constants and varied from one ply to another. From bending of thick and especially thin plate, the stresses along the thickness following quadratic law. The function $f_1(z)$ and $f_2(z)$ can be proposed then as:

$$f_1(z) = f_2(z) = \frac{1}{2}\left(\frac{h^2}{4} - z^2\right) \tag{12}$$

With this equation the contact force can be calculated and the result is in good correlation with test results for different stacking sequence, thickness and projectile diameter as shown in Fig. 21.

Other analytical model used mass-spring model [19]. The model is appropriate for impact velocity within the range where the equivalence quasi-static/dynamic is verified. The model of mass-spring can be seen in Fig. 22 where mass M_1 and M_2 correspond to projectile mass and effective mass of impacted plate. The spring coefficient k_c represents the stiffness of nonlinear contact, k_b bending stiffness and k_s shear stiffness of plate, k_m stiffness of membrane due to non-linearity of geometry. In general, mass M_2 is neglected and the stiffness k_m remains small and for the case of small displacement, all stiffness k_c, k_b, k_s in series are linear and can

Fig. 21 Comparison of contact force with different contact law

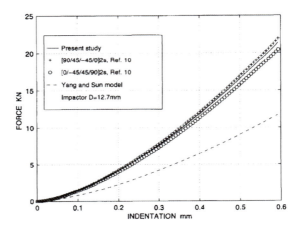

Fig. 22 Mass-Spring model of impact

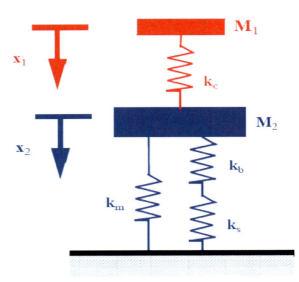

be replaced by one spring k. The integration of movement equation becomes simple and gave the expression of maximum contact force as:

$$P_{max} = V(kM_1)^2 = (2Uk)^{0.5} \qquad (13)$$

where U: impact energy and V: impact velocity.

The equation remains valid in the elastic domain of material. The proposed equations and analytical model allows modeling the impact globally for simple geometry of projectile. The equations did not allow analyzing completely the indentation and did not suit for identification of initial damage, nature and geometry of damage.

Fig. 23 Definition of direction in monolayer ply

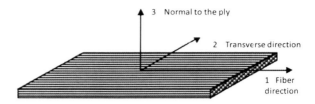

To simulate complex behavior of laminate plate including delamination, damage area and fiber failure involving the state of stress at each ply, numerical model such as finite element analysis is employed extensively. This FEA model will be explain later, but to be able to create the element and simulate the failure, criteria of matrix cracking, delamination and fiber fracture will be described in the next subchapter.

4 Failure Criteria and Failure Modeling of Laminate Composite

As mentioned previously that the damage mechanism of laminate subjected to impact loading in most of the case is started by matrix cracking followed by delamination and fiber fracture. This chapter describes the available mathematical model for the three failure modes as stated in different research papers. The numbering of directions for all criteria described in this sub-chapter is indicated in the Fig. 23:

4.1 Matrix Cracking

4.1.1 Matrix Cracking Criteria

Chang–Chang [23] has proposed for the case of tension loading in transverse direction ($\sigma_{22} > 0$), a criteria appearing only for in-plane state of stress. The criteria proposed that matrix cracking occurs if:

$$\left(\frac{\sigma_{22}}{Y_T}\right)^2 + \left(\frac{\tau_{12}}{S_{12}}\right)^2 \geq 1 \qquad (14)$$

where Y_T: normal stress strength of matrix in tension and in transverse direction.
  S_{12}: in-plane shear stress strength of matrix
The criteria did not take into account the shear stress τ_{23}, whereas it has a significant effect on matrix cracking produced by shear stresses. Hou [24] has proposed an improvement of criteria as the following:

$$\left(\frac{\sigma_{22}}{Y_T}\right)^2 + \left(\frac{\tau_{12}}{S_{12}}\right)^2 + \left(\frac{\tau_{23}}{S_{m23}}\right)^2 \geq 1 \tag{15}$$

where S_{m23} is the shear strength of matrix perpendicular to fiber direction.

Hashin [12] proposed also a criterion which distinguishes tension and compression normal stress. Moreover, the criteria introduced the contribution of normal stress σ_{33} acting perpendicularly to the laminate plane.

In the case of tension stress in transverse direction, matrix cracking is defined as the following:

$$\frac{(\sigma_{22} + \sigma_{33})^2}{\sigma_{MNT}^2} + \frac{(\tau_{23}^2 - \sigma_{22}\sigma_{33})}{\sigma_{MS}^2} + \frac{(\tau_{12}^2 + \tau_{13}^2)}{\sigma_{FS}^2} \geq 1 \tag{16}$$

For the case of compression stress in transverse direction:

$$\left(\left(\frac{\sigma_{MNC}}{2\sigma_{MS}}\right) - 1\right)\frac{(\sigma_{22} + \sigma_{33})}{\sigma_{MNC}} + \frac{(\sigma_{22} + \sigma_{33})^2}{4\sigma_{MS}^2} + \frac{\tau_{23}^2 - \sigma_{22}\sigma_{33}}{\tau_{MS}^2} + \frac{\tau_{12}^2 + \tau_{13}^2}{\tau_{FS}^2} \geq 1 \tag{17}$$

where MNT: normal strength of matrix in tension
MNC: normal strength of matrix in compression
σ_{MS} : shear strength of matrix
τ_{FS} : shear stress strength of fiber.

Another criterion has been proposed by Gosse and Mori [25] which is simpler in the form:

$$\frac{(\sigma_{22} + \sigma_{33})}{2Y} + \frac{1}{Y}\sqrt{\frac{(\sigma_{22} + \sigma_{33})^2}{4} + \tau_{23}^2} \geq 1 \tag{18}$$

where Y: normal stress strength in tension.

To conclude the long list of matrix cracking criteria, Tsai–Wu [26] proposed a general criterion in quadratic form as the following:

$$\frac{\sigma_{11}^2}{XX'} + \frac{\sigma_{22}^2 + \sigma_{33}^2}{YY'} + \frac{\tau_{12}^2 + \tau_{13}^2 + \tau_{23}^2}{S^2} + \frac{\sigma_{11}\sigma_{22} + \sigma_{11}\sigma_{33}}{\sqrt{XX'YY'}}$$
$$- \frac{\sigma_{22}\sigma_{33}}{YY'} + \sigma_{11}\left(\frac{1}{X} - \frac{1}{X'}\right) + (\sigma_{22} + \sigma_{33})\left(\frac{1}{Y} - \frac{1}{Y'}\right) \geq 1 \tag{19}$$

with X : normal stress strength in tension in direction 1
X' : normal stress strength in compression direction 1
Y : normal stress strength in tension in direction 2
Y' : normal stress strength in compression in direction 2
S : shear stress strength (supposedly the same in all direction)

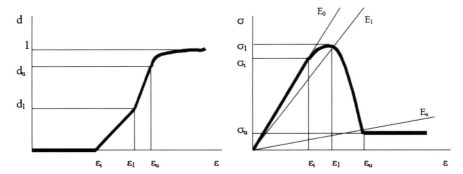

Fig. 24 Damage evolution law for matrix

4.1.2 Modeling of Matrix Cracking

Globally, modeling of matrix cracking was done by degrading certain mechanical properties on the elements. MAJEED [27] used Belytschko–Tsay beam with one integration point per ply for modeling a laminate plate $(-45/45/90)_{3s}$. Once Chang–chang criterion is fulfilled at the integration point of element, degradation of mechanical properties is imposed using $E_2 = G_{12} = \nu_1 = \nu_2 = 0$. It is a brutal degradation that would create problem during computation especially for high speed dynamic loading.

Hou [24] used eight points solid element. The laminate is modeled by one element per ply. The properties degradation inside plies is modeled using fracture mechanic. The degradation of E_1 (when $\sigma_{22} > 0$) and G_{12} are deteriorated by parameters which dependent only on time. Consequently, stresses σ_{22} and τ_{12} are decreasing and become zero at certain time. It is an improvement model compared to the previous one, however it is evident that the dependency of parameter to temporal variable alone would not satisfied.

The researches were led then to more complex modeling. McCarthy [28] used a bi-phase law which allows attributing different properties to fibers and matrix. It allows modeling the fiber fracture and matrix cracking separately. The stiffness matrix of ply is obtained by combining the stiffness of fiber and matrix:

$$C_{\text{pli}} = C_{\text{fiber}} \oplus C_{\text{matrix}} \quad (20)$$

The damage is controlled by the law as presented in Fig. 24. The FEA code (PAM-CRASH) utilized for the study allows calibrating the strain–stress curve of matrix. The strain values of ε_i, ε_1, ε_u is defined by its damage parameter corresponding to ε_1 and ε_u, noted as d_1 and d_u. To avoid the instability numerical computation, ε_u has been choose big enough from ε_1 in order to obtain non-brutal degradation. For $\varepsilon > \varepsilon_u$, residual strain non-zero is defined.

The proposed model was considered complex by Gauthier [29] taking into account significant number of parameters to be identified. Effectively, for matrix, it is necessary to determine the parameters for normal and shear stresses in tension and compression in the same time. In total, there are 30 parameters to be determined which need significant number of test to identify them.

4.2 Delamination

4.2.1 Delamination Criteria

The process of delamination from matrix cracking has been explained in the previous sub-chapter. In this sub-chapter different principal delamination criteria are presented which in most of the case in form quadratic form as for matrix cracking.

Hashin [12] proposed a criteria without taking into account the sign of σ_{33}, which leads to the limitation of its application for compression ($\sigma_{33} < 0$). The criterion is written as the following:

$$\left(\frac{\sigma_{33}}{\sigma_{DN}}\right)^2 + \frac{\tau_{23}^2 + \tau_{13}^2}{\tau_{DS}^2} \geq 1 \qquad (21)$$

where: σ_{DN}: normal strength of ply

τ_{DS}: delamination strength of ply due to shear

Brewer and Lagace [30] did not take into account either the sign of σ_{33} and proposed a similar criteria with Hashin and identical with criteria proposed by Chang and Springer [31] in the form :

$$\left(\frac{\sigma_{33}}{Z_T}\right)^2 + \frac{\tau_{23}^2}{S_{l23}^2} + \frac{\tau_{31}^2}{S_{31}^2} \geq 1 \qquad (22)$$

Where S_{13}: shear strength in plane 13

S_{l23}: delamination shear strength in plane 23

Z_T: normal strength along thickness direction.

Hou et al. [32] proposed an improvement criterion which takes into account the advantage of compression loading to withstand delamination and in the same time authorizing delamination produced by weak compression and high shear stress. The criterion takes into account then the sign of σ_{33}. The criterion distinguished three cases:

- When $\sigma_{33} \geq 0$. This case is the most probable to create delamination.

$$\left(\frac{\sigma_{33}}{Z_T}\right)^2 + \frac{\tau_{23}^2 + \tau_{13}^2}{(\delta + d_{ms}d_{fs})S_{13}^2} \geq 1 \qquad (22a)$$

- When $-\sqrt{\frac{\tau_{23}^2 + \tau_{13}^2}{8}} \leq \sigma_{33} \leq 0$. This is the case where even having compression stress, the delamination still occurs due to significant shear stress:

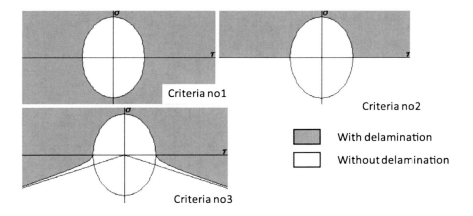

Fig. 25 Criteria evolution of Brewer and Lagace

$$\frac{\tau_{23}^2 + \tau_{13}^2 - 8\sigma_{33}^2}{(\delta + d_{ms}d_{fs})S_{13}^2} \geq 1 \qquad (23)$$

- When $-\sqrt{\frac{\tau_{23}^2+\tau_{13}^2}{8}} > \sigma_{33}$. There is no delamination.

With S_{13} : shear strength in plan 13

Z_T : normal strength in thickness direction

d_{ms} : damage coefficient of matrix in tension, varied from 0 to 1.

d_{fs} : damage coefficient of fiber fracture, varied from 0 to 1.

δ : ratio of interlaminate shear before and after appearance of damage in matrix or fiber.

Other research works improved the Brewer and Lagace criterion. The evolution of delamination criteria can be then summarize as the following:

- Criteria (no 1) without taking into account the advantage of compression stress σ_{33}.
- Criteria (no 2) which taking into account the advantage of compression stress σ_{33}
- Criteria (no 3) describing the possibility of delamination under presence of compression stress σ_{33} due to significant shear stress in transverse direction.

The 3 criteria are described in the Fig. 25 where the grey zone indicates the possible appearance of delamination.

Another simpler criterion has been proposed by Zhang [1] in the form:

$$\sqrt{\tau_{13}^2 + \tau_{23}^2} \geq ILSS \text{ or } \sigma_{33} > T_{3t} \qquad (24)$$

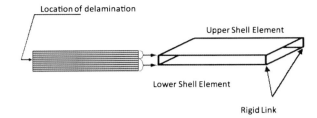

Fig. 26 Delamination modeling by Zhang

Where ILSS: interlaminate shear strength
T_{3t}: normal strength in the thickness direction

This criterion did not take into account the coupling between out-of-plane shear stress and tension stress which nevertheless represent the physical reality of phenomena found from the previous works.

4.2.2 Modeling of Delamination

The modeling of delamination found in the literature can be grouped into two categories:

- Plies separation using interface element

Each ply is meshed independently and two plies are joints by rigid link connecting the double nodes from each side of ply. The rigid connection is removed once the delamination criteria occur at the node or at the element of ply.

- Mechanical Properties degradation

Delamination is represented by mechanical properties degradation integrated in the stiffness matrix of the elements where delamination is detected.

Using Interface Elements Without Mechanical Properties Degradation

Zhang [1] proposed a simple model where it is assumed that the location of delamination is known. Half parts of laminate; upper and lower of interface where the delamination is supposedly occurred are modeled as shell elements (see Fig. 26). The elements are then joints by rigid link and delamination will occurs if the criteria set previously are fulfilled.

Different methods used to representing the link between plies were proposed by Fleming [33]. It was proposed using different type of element;

- Spring element with specific fracture criteria (Force based tied connection)
- element using force–displacement law including the degradation law (Cohesive Fracture model)

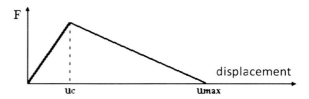

Fig. 27 Evolution of force as function of displacement

- and element using a law based on energy method where certain level energy propagates the initial delamination crack (Virtual Crack closure technique, VCCT).

The method of Force based tied connection used springs element joining nodes of 2 adjacent ply with the rupture criterion of the spring that can be written as the following:

$$\left(\frac{F_N}{F_{NC}}\right)^{a_n} + \left(\frac{F_S}{F_{Sc}}\right)^{a_s} \geq 1 \tag{25}$$

with F_N : normal force in the direction of rigid link

F_S : tangential force

F_{NC} : normal force at rupture

F_{Sc} : tangential force at rupture

a_n, a_s : parameters which govern interaction between modes of rupture.

However the force F_{NC} is difficult to be determined. The value is taken as $F_{NC} \sim \sigma_{ult}*A_e$ where A_e is the area around spring.

The cohesive fracture model used a force–displacement relation in the interface layer following certain form of law as presented in Fig. 27. The law avoids brutal separation between nodes, reduces numerical instability and allows dissipating the energy of system.

For virtual crack closure technique (VCCT), it is based on hypothesis that the necessary energy to propagate initial crack is equivalent to work done to bring the fissure back to the initial length. The nodes at the adjacent plies are joints using unidirectional elements. The rate of energy restitution in mode I can be written as:

$$G_I = \frac{1}{2\Delta A} F_I \left(u_{\sup} - u_{\inf}\right) \tag{26}$$

With F_I : interface force in mode I direction

u_{sup}, u_{inf} : nodes displacement in the adjacent plies in the vicinity of initial crack.

ΔA : increment of fissure surface which depends on the element size just beside the fissure.

The rate of restitution energy in mode II and III is written in the similar manner as

$$G_{II} = \frac{1}{2\Delta A} F_{II} \left(v_{\sup} - v_{\inf}\right) \tag{27}$$

$$G_{III} = \frac{1}{2\Delta A} F_{III} (w_{\sup} - w_{\inf}) \qquad (27a)$$

The failure criterion is proposed as the following:

$$\frac{G_I}{G_{Ic}} + \frac{G_{II}}{G_{IIc}} + \frac{G_{III}}{G_{IIIc}} = 1 \qquad (28)$$

The limitation of these methods relates by the fact that it simulates only the plies separation during delamination. The mechanical properties degradation during impact inside the plies is not taken into account. The degradation of plies, matrix cracking for example, seems important to be modeled to obtain a reasonable distribution of stress before delamination.

That is the reason, research works interested in developing more complex element taking into account the mechanical properties degradation and interface layer as well.

Using Mechanical Properties Degradation

The delamination is represented by reduction of mechanical properties inside plies. It creates the complex reduction of stresses related to the reduction of mechanical properties. Hou [32] modeled laminate composite by solid element (eight nodes) along the ply thickness. The mechanical properties are degraded progressively to zero to simulate the diminution of stress σ_{33} (for $\sigma_{33} > 0$), τ_{13} and τ_{23} in the element where the failure criteria is fulfilled. The comparison with experimental result is shown in Fig. 28 where the model correlates well with experimental result.

Another approach has been proposed by Bonini [34] where the laminate is modeled by 3D element using one element for each ply. The elements are created using double nodes in the interface which are subjected to a continuity condition of acceleration in 3 directions of space (see Fig. 29).

The delamination is presented by contact loss between two nodes in the interface. The condition of contact is imposed in such that after losing it, the two plies cannot be interconnected anymore.

Walrick [35] proposed a model using multi-layer shell element. A thin interface layer is introduced between plies prior to delamination. The failure due to delamination is presented with damage parameters d and d' which affect the stiffness matrix as the following:

- In the case of normal stress in transverse direction with cracking ($\sigma_{22} > 0$)

$$\begin{bmatrix} \sigma_{11} \\ \sigma_{22} \\ \tau_{12} \end{bmatrix} = \begin{bmatrix} \frac{E_{11}^0}{(1-\vartheta_{12}^0 \vartheta_{21}^0 (1-d'))} & \frac{\vartheta_{21}^0 E_{11}^0 (1-d')}{(1-\vartheta_{12}^0 \vartheta_{21}^0 (1-d'))} & 0 \\ \frac{\vartheta_{12}^0 E_{22}^0 (1-d')}{1-\vartheta_{12}^0 \vartheta_{21}^0 (1-d')} & \frac{E_{22}^0 (1-d')}{1-\vartheta_{12}^0 \vartheta_{21}^0 (1-d')} & 0 \\ 0 & 0 & G_{12}^0 (1-d) \end{bmatrix} \begin{bmatrix} \varepsilon_{11} \\ \varepsilon_{22} \\ \gamma_{12} \end{bmatrix} \qquad (29)$$

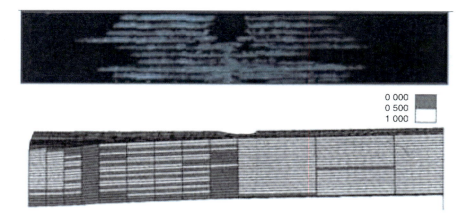

Fig. 28 Comparison experimental result and simulation using Hou model

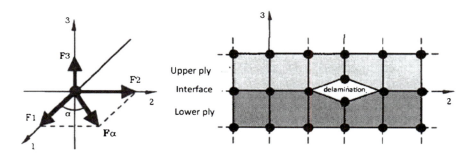

Fig. 29 Delamination model by Bonini

- In the case of $\sigma_{22} < 0$, only shear stress causes the failure:

$$\begin{bmatrix} \sigma_{11} \\ \sigma_{22} \\ \tau_{12} \end{bmatrix} = \begin{bmatrix} \frac{E_{11}^0}{(1-\vartheta_{12}^0\vartheta_{21}^0)} & \frac{\vartheta_{21}^0 E_{11}^0}{(1-\vartheta_{12}^0\vartheta_{21}^0)} & 0 \\ \frac{\vartheta_{12}^0 E_{22}^0}{1-\vartheta_{12}^0\vartheta_{21}^0} & \frac{E_{22}^0(1-d')}{1-\vartheta_{12}^0\vartheta_{12}^0} & 0 \\ 0 & 0 & G_{12}^0(1-d) \end{bmatrix} \begin{bmatrix} \varepsilon_{11} \\ \varepsilon_{22} \\ \gamma_{12} \end{bmatrix} \quad (30)$$

The parameters d and d' follow a degradation law described by Ladeveze and determined experimentally [36]. The detection of delamination is based on the evolution of energy density in the interface level that it can be expressed in function of stresses:

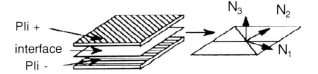

Fig. 30 Defintion of direction in the interface for Ladeveze et Allix model

$$\frac{dW}{d\vartheta} = \frac{1+\vartheta}{6E}\left[(\sigma_1 - \sigma_2)^2 + (\sigma_2 - \sigma_3)^2 + (\sigma_3 - \sigma_1)^2\right] + \frac{1-2\vartheta}{6E}(\sigma_1 + \sigma_2 + \sigma_3)^2 \quad (31)$$

The criterion is coupled with Tsai–Hill one. The delamination is translated by brutal increment of parameters value near to 1 when the drop of energy density reaches 20% and near to 0.1 when Tsai–Hill criterion is detected simultaneously. These critical values are determined empirically.

Ladeveze and Allix [36] used meso-model to simulate the behavior of laminate subjected to impact loading. The method is based on the irreversible process of thermodynamic theory where the state of material can be determined by knowing its certain internal parameters. For the case of isothermal elastic-damage, it is related to strain ε and damage parameters d_i.

The model is performed on the level meso-structure of laminate which means that the ply is considered homogenous along its thickness. The laminate is then defined by 2 components; shell and interface elements (see Fig. 30).

The shell undergoes the degradation due to micro-cracking of matrix, detachment fiber-matrix and fiber fracture while the interface simulates the delamination in mode I, II and III.

For non-damaged interface, deformation energy can be expressed as the following:

$$E_D = \frac{1}{2}\left(k_3^0(U_3)^2 + k_1^0(U_1)^2 + k_2^0(U_2)^2\right) = \frac{1}{2}\left(\frac{\sigma_{33}^2}{k_3^0} + \frac{\tau_{31}^2}{k_1^0} + \frac{\tau_{32}^2}{k_2^0}\right) \quad (32)$$

Where (U) is the different displacement between upper and lower surface of 2 adjacent plies with $(U) = U^+ - U^- = (U).N_1 + (U).N_2 + (U).N_3$ and k_1^0, k_2^0, k_3^0 are the initial stiffness elastic of interface ($k_1^0 = k_2^0 = k_3^0 = 0$ implicate the complete separation between surfaces). The relation between stresses and different displacement (U) is based on the orthotropic of interface and can be expressed as the following:

$$\begin{bmatrix} \tau_{13} \\ \tau_{23} \\ \sigma_{33} \end{bmatrix} = \begin{bmatrix} k_1^0 & 0 & 0 \\ 0 & k_2^0 & 0 \\ 0 & 0 & k_3^0 \end{bmatrix} \begin{bmatrix} U_1 \\ U_2 \\ U_3 \end{bmatrix} \quad (33)$$

The strain energy of damaged interface is expressed as the following:

$$E_D = \frac{1}{2}\left(\frac{\sigma_{33\sup}^2}{k_3^0(1-d_3)} + \frac{-\sigma_{33\sup}^2}{k_3^0} + \frac{\tau_{31}^2}{k_1^0(1-d_1)} + \frac{\tau_{32}^2}{k_2^0(1-d_2)}\right) \quad (34)$$

Where d_1, d_2, d_3 are the damage parameters associated to 3 modes of microcracking opening in the principle directions of interface. The thermodynamic forces associated to 3 damage parameters, and based on second law of thermodynamic, can be derived as follow:

$$Y_{d3} = \left.\frac{\partial E_D}{\partial d_3}\right|_{\sigma=cst} = \frac{1}{2}\frac{(\sigma_{33})_{\sup}^2}{k_3^0(1-d_3)^2} \text{ in Mode I}$$

$$Y_{d2} = \left.\frac{\partial E_D}{\partial d_2}\right|_{\sigma=cst} = \frac{1}{2}\frac{(\tau_{32})_{\sup}^2}{k_2^0(1-d_2)^2} \text{ in Mode II} \quad (35)$$

$$Y_{d1} = \left.\frac{\partial E_D}{\partial d_1}\right|_{\sigma=cst} = \frac{1}{2}\frac{(\tau_{31})_{\sup}^2}{k_1^0(1-d_1)^2} \text{ in Mode III}$$

The parameter which controls the damage evolution is defined as the following:

$$\underline{Y} = (\underline{Y}_{d3} + \gamma_1\underline{Y}_{d1} + \gamma_2\underline{Y}_{d2}) \text{ with } \underline{Y}_{di}|_t = \sup_{\tau \leq t} Y_{di}|_\tau \quad (36)$$

γ_1 and γ_2 are the coupling parameters and the 3 damage laws can be defined as:

$$d_3 = d_1 = d_2 = w(\underline{Y}) \text{ if } d3 < 1$$
$$d_3 = d_1 = d_2 = 0 \text{ otherwise} \quad (37)$$

with $w(Y) = \frac{\sqrt{Y}-\sqrt{Y_0}}{\sqrt{Y_c}-\sqrt{Y_0}}$ if $Y > Y_0$, $w(Y) = 0$ otherwise

Y_0 represents the threshold of damage energy, Yc for critical damage energy and again Yo, Yc, γ_1 and γ_2 are determined experimentally.

To improve the model for high velocity impact, Ladeveze proposed a delay effect which consists of describing the parameter evolution d with a certain delay in function of time. For a simple case of single direction, the evolution can be expressed as:

$$d = \frac{k}{a}[1 - \exp(-aw(Y) - d] \text{ if } d < 1$$
$$d = 0 \text{ otherwise} \quad (38)$$

k and a are the parameters characterizing the delay.

The model imposed a rate of damage maximum (k/a) depending on material used for laminate and consequently avoided the problem caused by severe impact producing too high damage speed and returning ineffective the delay effect. The advantage of the model is in prediction the damage phenomena and delamination. It is a possible model for simulation using plies by plies element or using multi-layer element.

4.3 Fiber Failure

The equation for fiber fracture criterion seems similar to delamination one. It is considered normal stress in fiber direction and in-plane shear stresses τ_{12} and τ_{13} for the failure criterion. The failure criteria proposed by Chang–Chang [23] is in the form:

$$e_f^2 = \left(\frac{\sigma_{11}}{X_T}\right)^2 + \left(\frac{\tau_{12}}{S_{12}}\right)^2 \geq 1 \tag{39}$$

With X_T : normal stress strength in fiber direction
S_{12} : In-plane shear stress strength

Hou [24] proposed a modification on the criterion by taking into account the shear stress τ_{13} having the same effect as τ_{12}. It should be noted that the shear strength for fiber fracture is taken independently from matrix one. The modification law proposed by Hou is expressed as the following:

$$e_f^2 = \left(\frac{\sigma_{11}}{X_T}\right)^2 + \left(\frac{\tau_{12}^2 + \tau_{13}^2}{S_f^2}\right)^2 \geq 1 \tag{40}$$

with S_f : shear strength of fiber.

The modeling of fiber fracture follows the same reflection as for delamination which has been explained in the previous sub-chapter. The damage of fiber followed an evolution of mechanical properties degradation as for in the delamination (d_i) with certain law. The criterion is expressed in term of stress, it could be interesting to find out a criterion based on strain which would be simpler to be implemented for numerical analysis.

5 Discussion and Conclusion

In this chapter the experimental set up and method of impact testing has been detailed. The method to obtain the contact force and energy absorption by the specimens seems becoming a standard method for this type of testing. It can be used for different type of specimen since all calculation is derived from history of force given by load cell and initial velocity obtained from displacement sensor using laser or optical system located in the contact surface level.

The analyzing of experimental results in term of velocity, force and energy absorption during impact explains the damage mechanism in term of threshold of initiation damage, delamination and fiber fracture. The absorption energy capability of laminate plate can be determined as well.

The effect of parameters such as boundary conditions, projectile shape, laminate thickness, fiber direction and others has been highlighted. The study on the effect of parameters gave a qualitative tendency on energy absorption and contact force which seemingly would be different for different material.

The failure criteria have been also detailed form the simplest until the latest one. The highlighted failure criteria includes for matrix cracking, delamination and fiber fracture. The model to simulate the laminate failure subjected to impact loading consists of proposing an element with failure criteria integrated to it as for matrix cracking. For delamination, there are 2 proposed models; without interface layer and with degradation of mechanical properties. The interface layer element integrates the degradation mechanical properties law where at certain values the layer will be separated. The parameters involves in the degradation law has to be determined experimentaly. A complete model to simulate delamination has been proposed by Ladeveze where the matrix cracking is simulated by degradation at the ply element to give a real stress distribution before delamination, and delamination is simulated by interface layer using simulation at meso-structure level. The failure criteria and its element model which have been proposed in this study become the basis for simulation using finite elements analysis and are integrated to the elements definition in the latest version of FEA software.

Finally, the aim of the simulation is to be simple and robust to be used for industrial application where the structure becomes complex and needs high number of elements to be run. High number of elements with non-linear law of failure criteria would lead to divergence of computation. Therefore a simple failure criteria and its related elements to simulate all three failures; matrix cracking, delamination and fiber fracture still need to be investigated. An array of spring located at double nodes at the adjacent plies, for example, seems a good candidate to simulate the delamination where its force–displacement law can be determined using experiment of fracture mechanics. The development of special element where a solid element can be divided into two element of shell once the delamination criterion is fulfilled, becomes also a direction of research in FEA area.

References

1. Zhang, X., Davies, G.A.O., Hitchings, D.: Impact damage with compressive preload and post-impact compression of carbon composite plates. Int. J. Impact Eng. **22**, 485–509 (1999)
2. Rivallant, S.: Modelisation a l'impact de pales d'helicopteres. These de Doctorat, SupAero (2003)
3. Aminanda, Y.:Contribution a l'analyse et a la modelisation de structures sandwiches impactees.2. These de Doctorat (2004)
4. Tsotis, T.K., Lee, S.M.: Characterization of localised failures modes in honeycomb sandwich panels using indentation. ASTM STP **1274**, 139–165 (1996)

5. Hallett, S.R.: Three points beam impact test on T300/914 carbon-fibre composites. J. Compos. Sci. Technol. **60**, 115–124 (2000)
6. Thevenet, P.: Impact sur structure sandwich. Rapport interne DTP Toldom 2000, EADS CCR (1998)
7. Herup, E.J., Palazotto, A.N.: Low-velocity impact damage initiation in graphite/epoxy/nomex honeycomb-sandwich plates. Compos. Sci. Technol. **57**, 1581–1598 (1997)
8. Gay, D.: Matériaux Composites, 3ème Edition revue et augmentée, Paris Hermès (1991)
9. Richardson, M.O.W., Wisheart, M.J.: Review of low-velocity impact properties of composite materials. Compos. Part A **27A**, 1123–1131 (1996)
10. Bonini, J.: Contribution à la prédiction numérique de l'endommagement de stratifies composites sous impact basse vitesse. Thèse de doctorat, ENSAM Bordeaux (1995)
11. Chang, F.K., Chang, K.Y.: A progressive damage model for laminate composites containing stress concentrations. J. Compos. Mater., Sept. 834–855 (1987)
12. Hashin, Z.: Failure criteria for unidirectional fiber composites. J. Appl. Mech. **47**, 329–334 (1980)
13. Bernard, M.L., Lagace, P.A.: Impact resistance of composite sandwich plates. J Reinforced Plastics compos **8**(9):432–445 (1989)
14. Tsang, P.W., Lagace, P.A.: Failure mechanisme of impact-damaged sandwich panels under uniaxial compression. Proc of the 35th AIAA/ASME/ASCE/AHS/ASC struc Dyn and matconference 2:745–754 (1994)
15. Pillot, S.: Damage mechanism of carbon/epoxy and jute/epoxy laminate plate. Repport of project. (2009)
16. MitrevskiI, T., Marshall, I.H.,Thomson, R., Jones, R., Whittingham, B.: The effect of impactor shape on the impact response of composite laminates Compos. Struct. (2004)
17. Edgar Fuossa, T., Paul V. Straznickya, Cheung Poonb.: Effects of stacking sequence on the impact resistance laminates. Part 2: prediction method, J. Compos. Struct. 41 177–186 (1998)
18. Belingardi, G., Vadori, R.: Influence of the laminate thickness in low velocity impact behavior of composite material plate. J. Compos. Struct. **61**, 27–38 (2003)
19. Abrate, S.: Localised impact on sandwich structures with laminated facings. Appl. Mech Rev **50**(2), 69–82 (1997)
20. Wang, C.Y., Yew, C.H.: Impact damage in composite laminates. Comput. Struct. **37**(6), 967–982 (1990)
21. Choi I.H., Lim C.H.: Low-velocity impact analysis of composite laminates using linearized contact law, J. Compos Struct. (2004)
22. Aboussaleh M., Boukhili, R.: The contact behavior between laminated composites and rigid impactors. J. Compos. Struct. **43**, 165–178 (1998)
23. Chang, F.K., Chang, K.Y.: A progressive damage model for laminate composites containing stress concentrations. J. Compos. Mater., 834–855 Sep. (1987)
24. Hou, J.P., Petrinic, N., Ruiz, C., Hallett, S.R.: Prediction of impact damage in composite plates. Compos. Sci. Technol. **60**, 273–281 (2000)
25. Finn, S.R., Springer, G.S.: Delaminations in composite plates under transverse static or impact loads - a model. Compos. Struct. **23**(3), 177–190 (1993)
26. Tsai, S.W., Wu, E.M.: A general theory of strength for anisotropic materials. J. Compos. Mater. 58–80 Jan. (1971)
27. Majeed, O., Worswick, M.J., Strazicky, P.V., Poon, C.: Numerical modeling of transverse impact on composite coupons. Canadian Aeronautics and Space Journal **40**, 99–106 (1994)
28. McCarthy, M., Harte, C., Wiggenraad, J., Michielsen, A., Kohlgrueber, D.: Finite element modelling of crash response of composite aerospace sub-floor structures. Comp. Mech. **26**, 250–258 (2000)
29. Gauthier, C.: Contribution à la modélisation du comportement en crash des structures stratifiées métal/composite : développement d'un élément de coque multicouches multimatériaux. Thèse de doctorat, Université de Valenciennes et du Hainaut-Cambrésis, (1996)

30. Brewer, J.C., Lagace, P.A.: Quadratic stress criterion for initiation of delamination. J. Compos. Mater. **22**(12), 1141–1155 (1988)
31. Banerjee, R.: Numerical simulation of impact damage in composite laminates. In: Proceedings of the 7th Technical Conference of the American Society for Composites: 539–552 (1992)
32. Hou, J.P., Petrinic, N., Ruiz, C.: A delamination criterion for laminated composites under low-velocity impact. Compos. Sci. Technol. **61**, 2069–2074 (2001)
33. Fleming, D.C.: Delamination modeling of composites for improved crash analysis. NASA 209725 (1999)
34. Bonini, J.:Contribution à la prédiction numérique de l'endommagement de stratifies composites sous impact basse vitesse. Thèse de doctorat, ENSAM Bordeaux (1995)
35. Walrick, J.C.: Contribution au développement d'une nouvelle méthodologie pour l'étude du délaminage dans les structures stratifiées composites : application à l'impact basse vitesse. Thèse de doctorat, Université de Valenciennes et du Hainaut-Cambrésis (1999)
36. Allix, O., Ladeveze, P., Corigliano, A.: Damage analysis of interlaminar fracture specimens. Compos. Struct. **31**, 61–74 (1995)

The Blast Response of Sandwich Structures

M. Yazid Yahya, W. J. Cantwell, G. S. Langdon and G. N. Nurick

Abstract This project studied the response of sandwich panel subjected to blast loading. The panels were based on two different face sheets (aluminium and woven glass-fibre/epoxy) and an aluminium honeycomb core. Experimental studies were carried out to analyse the effect of skin and core thickness on the blast response of the panels. The sandwich panels with glass-fibre/epoxy face sheets exhibited delamination in the face skin and core crushing, whereas failure in the sandwich panels with aluminium skins involved permanent visible indentation and core crushing. It was concluded that the composite-skinned sandwich structures offered a superior blast resistance to the aluminium-skinned system.

Keywords Sandwich panels · Blast · Honeycomb · Composite

1 Introduction

Composite materials, such as sandwich structures are finding increasing use in a number of primary aircraft structures. One particularly interesting development is the use of sandwich structures in the manufacture of the fuselage of the Boeing

---

M. Y. Yahya (✉)
Center for Composites, Universiti Teknologi Malaysia,
81310 Skudai, Johor, Malaysia
e-mail: yazid@fkm.utm.my

W. J. Cantwell
Department of Engineering, Impact Research Centre,
University of Liverpool, Brownlow Hill,
Liverpool L69 3GH, UK

G. S. Langdon · G. N. Nurick
Blast Impact and Survivability Research Unit (BISRU),
Department of Mechanical Engineering, University of Cape Town,
Private Bag, Rondebosch 7701, South Africa

787 Dreamliner. In recent years there have been a number of instances of explosions or blasts occurring within pressurized aircraft fuselages, with one of the most recent being the bursting of an oxygen cylinder in a Qantas Airlines aircraft in July 2008. However, in spite of the fact that the dynamic response of composites has been investigated by a large number of workers, relatively little work has been conducted to investigate the blast response of sandwich structures.

Zu et al. [1] investigated the effects of different face-sheet and core configurations on structural response of sandwich structures, i.e. face-sheet thickness, cell size and foil thickness of the honeycomb. They found that specimens with thicker face-sheets, a higher density core and loaded by larger charges tended to exhibit localised deformation on the front face and those with thinner skins and a low density core and subjected to lower level blasts were prone to deform globally. Based on a quantitative analysis, it was also been found that the face-sheet thickness and relative density of core structure can significantly affect the back face deformation. It was evident that the back face deflection increased with impulse in an, approximately linear fashion.

For blast resistant applications, Hansen et al. [2] investigated the response of aluminium foam, on a rigid back plate, to close range explosions. Foam offered the potential to absorb the impulse arising from the relatively short blast duration, high pressure shock front and modify it for transmission through the foam (or in fact, any cellular material) into a longer duration, lower magnitude force. This offers potential for controlled energy absorption and reduced force transfer compared to equivalent solid plates, although the mechanisms of shock transfer are still not fully understood.

Karagiozova et al. [3] presented an experimental and numerical investigation into the response of flexible sandwich-type panels subjected to blast loading. The response of sandwich-type panels with steel plates and polystyrene cores were compared to panels with steel plates and aluminium honeycomb cores. The panels were loaded by detonating plastic explosive discs in close proximity to the front face of the panel. The numerical model was used to explain the stress attenuation and enhancement of the panels with different cores, when subjected to blast-induced dynamic loading. The permanent deflection of the back plate was determined by the velocity attenuation properties (and hence the transmitted stress pulse) of the core. Core efficiency in terms of energy absorption is an important factor for thicker cores. For panels of comparable mass, the aluminium honeycomb cores performed better than those with polystyrene cores.

McKown et al. [4] investigated the behaviour of lattice structures under blast loading. The blast resistance of the lattice structures increased with increasing yield stress and was shown to be related to the structures specific energy-absorbing characteristics.

Radford et al. [5] studied the dynamic responses of clamped circular monolithic and sandwich plates of equal areal mass by loading the plates at their mid-span by metal foam projectiles. The sandwich plates comprised stainless steel face sheets and aluminium alloy metal foam cores. It is found that the sandwich plates offer a higher shock resistance than monolithic plates of equal mass. Further, the shock resistance of the sandwich plates increased with increasing thickness of sandwich core. Finite element simulations of these experiments were in good agreement with the experimental measurements and demonstrated that the strain-rate sensitivity of stainless steel plays a significant role in increasing the shock resistance of the monolithic and sandwich plates.

Fleck and Deshpande [6] proposed an analytical model for predicting the finite deflection response of clamped sandwich beams subjected to shock loading, including the effects of fluid–structure interaction. They demonstrated the accuracy of their analytical model in the case of no fluid-interaction, by direct comparison with the finite element calculations of Xue and Hutchinson [7] for clamped sandwich beams.

Sriram and Vaida [8] modelled aluminium foam sandwich composites subjected to blast loads using LS-DYNA software. The sandwich structure was manufactured using laminated face sheets and an aluminium core. Damage progression in the sandwich occurred by 'dishing', which increased with increasing severity of the blast.

It is clear that there are still many unanswered questions regarding the response of sandwich panels to blast loading. In this paper, the results of blast loading experiments on two types of sandwich panel are presented. The panels were all based on the same honeycomb core but had differing face-sheet materials (glass-fibre reinforced epoxy or aluminium alloy) and core heights (13 and 25 mm). The materials were chosen on the basis that they were commercially available from the same company (Hexcel) and thus the core material could be assumed to be consistent across the test range.

2 Experimental Procedure

The aluminium honeycomb sandwich structures examined in this study, Aeroweb 3003, were supplied in the form of large flat panels by Hexcel Ltd. The 0.6 mm thick composite skins in the sandwich structures were based on a woven glass-fibre reinforced epoxy with a fibre volume fraction of 55% [9]. The density of the aluminium honeycomb was 84 kg/m^3 and the cell size (face to face) was approximately 6 mm. Two thickness of aluminium honeycomb core were investigated, these being 13 and 25 mm. The Aeroweb 3003 panels are a lightweight, high performance structural sandwich panel. The panel exhibits superior mechanical and physical properties, it is easy to install and can be readily cut and machined in the laboratory or workshop. A summary of the key mechanical and physical properties of these panels is given in Table 1. The panels were cut using a diamond circular saw to minimise the amount of damage incurred during the preparation process.

The second sandwich panel investigated in this study was based on an aluminium alloy honeycomb core and aluminium alloy face sheets. Core thicknesses

Table 1 Mechanical properties of the Aeroweb 3003 aluminium honeycomb core [9]

Property	Specification
Density (kg/m^3)	84
Cell size, l (mm)	6.4
Foil thickness, t (mm)	0.064
E_{11} (MPa)	0.947
V_{12}	0.3

Table 2 Mechanical properties of the aluminium honeycomb core [10]

Core thickness (mm)	Cell size (mm)	Foil thickness (mm)	Density (kg/m^3)
13	5.7	0.080	96
30	6.4	0.064	84

of 13 and 30 mm were used in this study. A summary of the physical properties of the honeycomb is given in Table 2.

3 Ballistic Pendulum

A ballistic pendulum was used to measure the impulse imparted to the test plate. The ballistic pendulum consisted of a steel I-beam suspended on four spring steel cables as shown in Fig. 1. The spring steel cables were attached to the I-beam of the ballistic pendulum by four adjustable screws. The pendulum was levelled by adjusting the screws and verified using a spirit-level. Counter-balancing masses were attached at one end of the I-beam. The I-beam balancing masses are used to counter the mass of the test rig attached on the other end of the ballistic pendulum, ensuring that all four spring steel cables carry the same load. The impulse generated by the explosion is then transmitted through the centroid of the pendulum. A soft tipped recording pen was attached to the pendulum at the same end as the counter-balancing masses, to record the oscillation amplitude of the pendulum on a sheet of tracing paper. The oscillation is directly related to the impulse transmitted to the test specimen.

In order to calculate the impulse from the tracing paper, several measurements need to be taken off the apparatus.

4 Compression Properties of the Sandwich Structures

Typical force–displacement curves following compression tests on the aluminium skin/honeycomb core sandwich structures (13 and 30 mm thick cores) and the glass-fibre/epoxy skin honeycomb core sandwich structures (13 and 25 mm thick

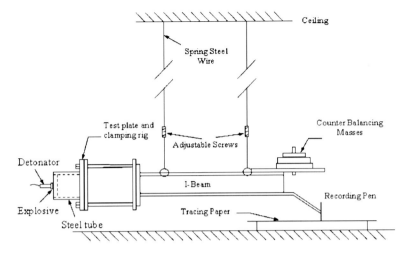

Fig. 1 Schematic of the ballistic pendulum

cores) are shown in Figs. 2 and 3. From the figures, it is apparent that all the graphs exhibit a linear load–displacement trace until the maximum load is reached. After the peak load, a sharp drop is observed. These curves agree with those of Aminanda et al. [11] and Othman and Barton [12] following tests on woven carbon skin/honeycomb sandwich structures. During testing, a folding mechanism started roughly in the middle of the sample as can be seen in Fig. 4. This is in contrast to the previously reported findings by Wu and Jiang [13], where the buckling of the aluminium honeycomb core was found to initiate from the upper surface and move progressively downwards.

The average values of compressive strength for the 13 mm thick aluminium skinned system were 4.68 and 4.04 MPa for the 30 mm thickness system. In contrast, the values of compressive strength for the glass-fibre skin honeycomb core are quite similar for both thicknesses, those being 3.16 (13 mm thick) and 3.12 MPa (25 mm thick).

Paik et al. [14] conducted crushing tests on aluminium honeycomb sandwich panel specimens, varying the cell thickness and height of the honeycomb core. They showed that the core height is not a significant parameter in the crushing behaviour of the honeycomb core. As would be expected, however, the wall thickness of the honeycomb cell has a more pronounced effect on the crush strength of sandwich panels subjected to longitudinal load.

5 The Blast Response of the Sandwich Structures

A total of 23 experiments were conducted to evaluate the structural response of sandwich panels under blast loading. Charges with four different mass were used; 1.5, 2.0, 2.5 and 3.0 g, to produce different levels of impulses. The impulses,

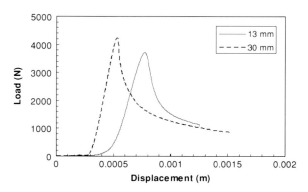

Fig. 2 Typical force–displacement curves following compression test on the aluminium/honeycomb with thicknesses of 13 and 30 mm

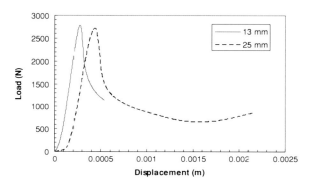

Fig. 3 Typical force–displacement curves following compression test on the glass-fibre epoxy/honeycomb with thicknesses of 13 and 25 mm

deflections and failure modes for the sandwich panels after testing are given in Table 3.

6 Experimental Observations

Based on the configuration of the sandwich panel, the deformation and failure modes of specimens observed following the tests can be classified with respect to the front face sheet, core and back face sheet.

Figures 5, 6, 7, 8, 9, 10, 11, 12 show the front and back faces of the sandwich panels after testing and optical micrographs of the damaged specimens. The permanently deformed profiles differ according to the distance between the plastic explosive and the test specimen. The back face sandwich profile resembled a uniform dome shape, as shown in Figs. 5, 7, 9 and 11. The profiles are similar to those reported by Teeling-Smith and Nurick [16] following blast tests on uniformly-loaded circular steel plates. The load distribution is assumed to be uniform over the panel area for stand-off distances between 90 and 180 mm. For lower stand-off distances (i.e. between 45 and 65 mm), the plate resembles a smaller inner dome superimposed on a larger global dome, as shown in Figs. 5a and 7a.

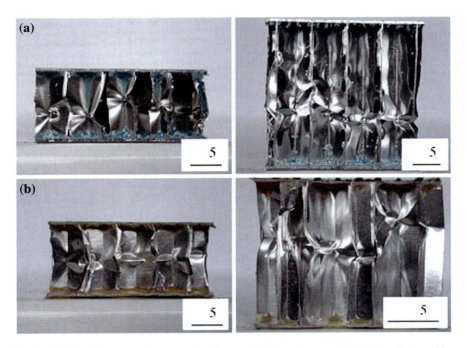

Fig. 4 Deformed compression samples for sandwich structures with (**a**) aluminium skins (**b**) glass-fibre/epoxy skins

This deformation profile concurs with the experimental observations reported by Nurick and Martin [17] for circular plates loaded using disc-shaped plastic explosive mounted directly on the test plate (i.e. subjected to localised blast loading). The load distribution is considered to be localised to the centre of the plate. It was found that both thicknesses exhibited similar failure modes.

6.1 Aluminium Skin/Honeycomb Core Sandwich Structure With a 13 mm Thick Core

Figure 5 shows photographs of the front and back faces of panels following impulse between 2.4 and 4.6 N s. All of the specimens exhibit a uniform dome shape on the back face. The front face suffered a localised indention failure and global deformation shape at an impulse 2.4 N s. Global deformation increased with increasing impulse between 1.7 and 4.6 N s. The specimens subjected to impulses of 2.4 and 2.7 N s showed typical Mode I (large inelastic deformation failure) and the plate at impulse of 4.6 N s exhibited a Mode II* failure (large inelastic deformation with tensile tearing at the end).

Table 3 Blast tests results on the sandwich structures, sorted by skin type, panel thickness, charge mass and stand-off distance (SOD)

Skin	Specimen	Charge Mass g	SOD mm	Panel thickness mm	Impulse Ns	Deflection (mm) Front	Failure Back	Failure Mode
Aluminium alloy	J1	1.5	45	13.90	2.4	13.16	6.23	
	J2	2.0	90	13.74	2.7	9.10	6.77	
	J3	2.5	180	13.80	4.6	19.55	–	II* (lift)
	K1	2.0	45	29.80	2.8	–	6.30	II (front)
	K2	2.0	90	29.72	3.1	13.14	4.69	
	K3	3.0	90	29.80	4.7	20.76	9.56	II*(back)
	K4	3.0	180	29.90	5.3	25.75	11.18	II*(back)
Composite	L1	1.5	45	13.29	2.4	–	4.52	II (Front)
	L2	1.5	90	13.55	2.8	6.18	4.74	
	L3	1.5	180	13.60	2.5	3.47	3.39	
	L4	2.0	90	13.60	3.3	9.34	7.29	
	L5	2.0	180	13.60	3.4	10.40	7.49	
	L6	2.5	180	13.00	4.7	13.47	13.71	II*(back)
	M1	1.5	45	26.00	2.3	5.62	3.28	
	M2	1.5	90	26.14	1.7	0.61	0.38	
	M3	1.6	90	26.00	2.7	3.90	2.27	
	M4	1.5	180	26.10	3.1	3.47	2.03	
	M5	2.0	45	26.22	2.9	–	5.20	II (front)
	M6	2.0	65	25.90	3.3	7.09	4.59	
	M7	2.0	90	26.20	3.8	9.04	4.64	
	M8	2.0	180	26.22	3.4	7.69	4.71	
	M9	3	90	25.90	4.9	14.23	9.39	
	M10	3	180	25.72	4.9	14.53	10.12	

- Failure Mode II – Complete tearing Mode II* - Large inelastic deformation with partial tearing around part of the boundary Mode I failure is defined as large inelastic deformation. Mode II as complete tearing and the transition between the two failure modes is defined as Mode II* [15]

Figure 6 shows typical cross-sections of the 13 mm thick aluminium skinned honeycomb specimen after blast testing. Following an impulse of 2.4 N s, localised buckling of some of core members is apparent in the lower segment, whereas the upper segment on the cell wall remain straight and underformed. In Fig. 6b, the honeycomb core is only partially crushed at the lowest impulse of 2.7 N s. At the highest impulse of 4.6 N s, core debonding from the face skin was observed, resulting in localised separation of the skin and core.

6.2 Aluminium Skinned Honeycomb Structures

Figure 7 shows photographs for the front and back faces of the aluminium skinned honeycomb core with a thickness of 30 mm, subjected to impulses between 2.8 and 5.3 N s. Pitting damage on the front face and a small bulge occurs at the centre of the back face in the panel subjected to an impulse 2.8 N s (localised loading).

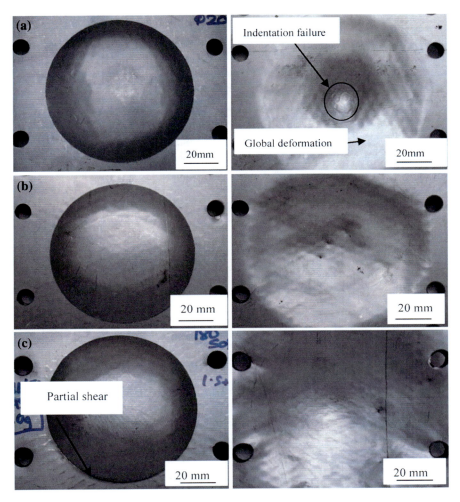

Fig. 5 Front and back surfaces of the 13 mm thick aluminium skinned sandwich structures after blast testing (**a**) Impulse = 2.4 N s (Panel J1) (**b**) Impulse = 2.7 N s (Panel J2) (**c**) Impulse = 4.6 N s (Panel J3)

The back face deformation profiles for all the panels have the shape of a uniform dome. The global deformation mode increased with increasing impulse between 3.1 and 4.7 N s. Following an impulse of 5.3 N s, the front face skin exhibits a crack in the centre and partial shear at the back face.

Figure 8 shows cross-sections of the specimens following testing. The specimen tested at an impulse 2.8 N s shows extensive core crushing through the thickness in the centre and shear failure on the front and back faces. The deformed honeycomb core shows a progressive deformation pattern, which is the similar to that observed following low-velocity impact experiments [14]. At impulses

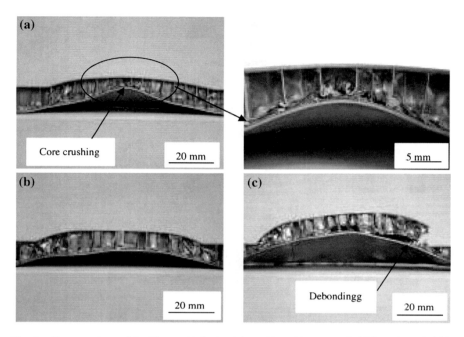

Fig. 6 Cross-sections of the 13 mm thick aluminium skinned honeycomb (a) Impulse = 2.4 N s (b) Impulse = 2.7 Ns (c) Impulse = 4.6 N s

between 3.1 and 5.3 N s, progressive buckling is localised to the side adjacent to the front face, and the cell walls remain virtually straight. This is consistent with previous finding on the dynamic crush behaviour of square honeycomb sandwich cores by Xue and Hutchinson [18]. Core crushing in these samples increases with increasing impulse. Shear failure between the back skin and core occurred at impulses of 4.7 and 5.3 N s.

6.3 Glass-Fibre Epoxy/Honeycomb Core Sandwich Structures With a 13 mm Thick Core

Figure 9 shows photographs of the 13 mm thick glass-fibre/epoxy sandwich structure. No damage on the front and back skins was observed in the samples subjected to 2.5, 2.8, 3.3 and 3.4 N s. All of the panels show a dome-shaped back surface deformation. The panel subjected to an impulse of 4.7 N s exhibited partial shear at the back surface. Pitting failure occurred at the front face of the panel subjected to impulse 2.4 N s (localised loading).

Figure 10 shows cross-sections of these sandwich panels. Damage in the honeycomb core took the form of localised buckling in a cell wall, increasing from 2.8 to 4.7 N s. In the case of localised loading (see Fig. 10a), it is apparent that the

The Blast Response of Sandwich Structures

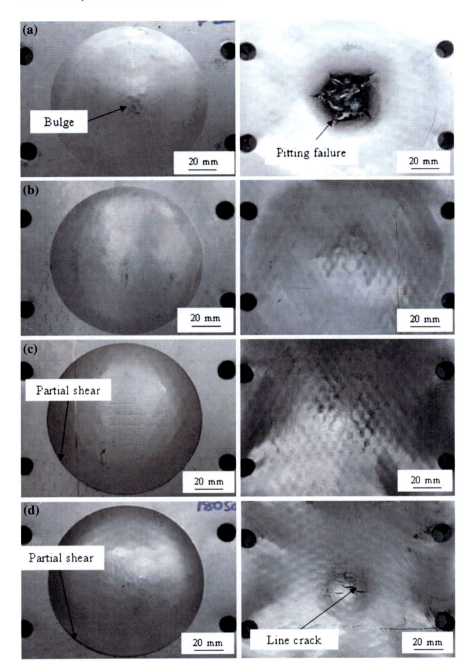

Fig. 7 Front and back surfaces of the 30 mm thick aluminium skinned sandwich structures after blast testing (**a**) Impulse = 2.8 N s (Panel K1) (**b**) Impulse = 3.1 N s (Panel K2) (**c**) Impulse = 4.7 N s (Panel K3) (**d**) Impulse = 5.3 N s (Panel K4)

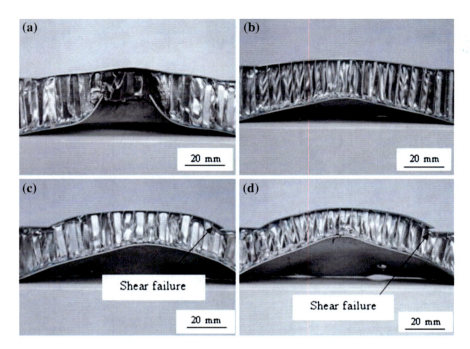

Fig. 8 Cross-sections view of the 30 mm thick aluminium skinned honeycomb sandwich structures (**a**) Impulse = 2.8 N s (**b**) Impulse = 3.1 N s (**c**) Impulse = 4.7 N s (**d**) Impulse = 5.3 N s

front skin has been ruptured and the aluminium core is heavily crushed. Following an impulse of 4.7 N s, delamination and debonding occurred between the front face and the core.

6.4 Glass-Fibre/Epoxy Skinned Honeycomb Structures

Figure 11 shows photographs of the front and back faces of the 25 mm thick glass-fibre/epoxy skinned honeycomb core panels subjected to impulses between 1.7 and 4.9 N s. For localised loading (the stand-off distance was 45 mm), a global deformation mode was apparent following an impulse of 2.3 N s. Pitting damage of the front face occurs when the impulse is increased to 2.9 N s, whereas when the impulse increases to 3.3 N s (stand-off distance was 65 mm), the front face exhibited global deformation but was not damaged.

Following uniform loading at an impulse of 1.7 N s, the specimen showed a small localised indentation failure on the front face. The global deformation mode increased with increasing impulse between 2.7 and 4.9 N s (see Figs. (11d–j).

Figure 12 shows sections of these specimens following blast loading. Core crushing in the cell walls was apparent in the honeycomb core of the glass-fibre/

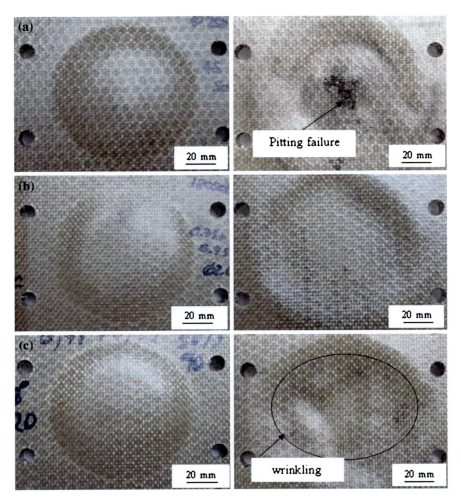

Fig. 9 Front and back surfaces of the glass-fibre epoxy/aluminium honeycomb after blast testing (core thickness = 13 mm) (**a**) Impulse = 2.4 N s (Panel L1) (**b**) Impulse = 2.5 N s (Panel L3) (**c**) Impulse = 2.8 N s (Panel L2) (**d**) Impulse = 3.3 N s (Panel L4) (**e**) Impulse = 3.4 N s (Panel L5) (**f**) Impulse = 4.7 N s (Panel L6)

epoxy sandwich specimens directly under the point of loading as seen in Figs. 11a–c. The glass-fibre skin ruptured at an impulse of 2.9 N s but not at impulses of 2.3 and 3.3 N s (only delamination/debonding between the face sheet and the core were observed). Damage following an impulse of 2.3 N s consists of top surface fibre fracture and localised core crushing. The sandwich panels subjected to an impulse of 2.9 N s exhibited core damage that extended through the thickness of the core.

During uniform loading, plastic deformation and core crushing in the lowermost region of the sandwich structure increases with increasing impulse between

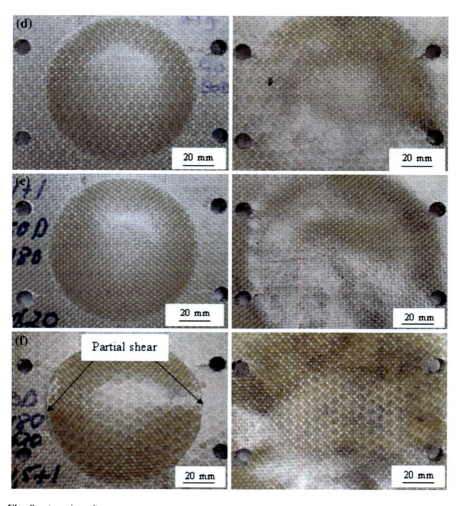

Fig. 9 (continued)

1.7 and 4.9 N s. The crushing strains are greatest in the central core members as a result of the greatest applied pressure associated with blast loading [19]. Debonding and delamination between the core and the front skin occurs between 3.8 and 4.9 N s. A small tearing failure on the back face initiates at the boundary of the sandwich structure is apparent after 4.9 N s (Fig. 11i). The bottom and front face sheets were left undamaged at impulses between 1.7 and 3.8 N s.

The failure characteristics of the sandwich structure are also influenced by core thickness. Mode II* failure in the thin sandwich structure occurs at an impulse of 4.7 N s (see Fig. 12f) whereas for the thick sandwich with a composite skin, this occurs at an impulse of 4.9 N s (see Fig. 12j).

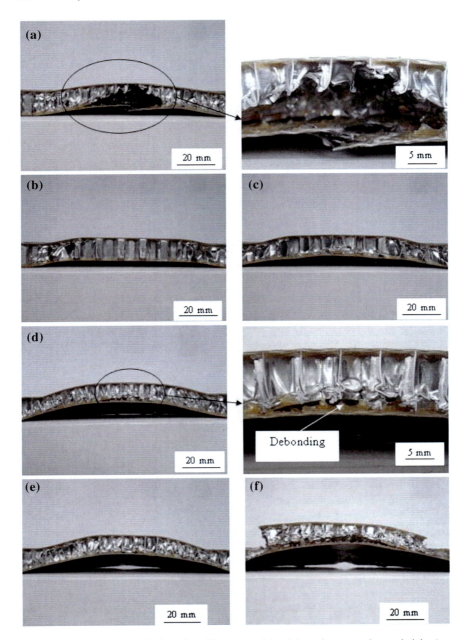

Fig. 10 Cross-sections of the glass-fibre epoxy/aluminium honeycomb sandwich (core thickness = 13 mm) (**a**) Impulse = 2.4 N s (**b**) Impulse = 2.5 N s (**c**) Impulse = 2.8 N s (**d**) Impulse = 3.3 N s (**e**) Impulse = 3.4 N s (**f**) Impulse = 4.7 N s

7 Quantification of the Damage Within the Sandwich Panels

In this investigation, the failure characteristic of the sandwich panels was significantly different from the conventional laminated structures [20, 21] examined in being strongly dependent on the characteristics of the core and skin materials.

Figures 13, 14, 15 and 16 show the variation of the permanent front and back face displacement with impulse for the aluminium-skinned and the glass/epoxy-skinned sandwich structures. In most cases, the mid-point deflection increased linearly with increasing impulse. The front face displacement was generally greater than the back face displacement. The deflection of the aluminium skinned sandwich structures for both thicknesses was slightly greater than that of the glass/epoxy skinned sandwich structures, suggesting that the glass-fibre/epoxy sandwich panels offer a superior blast performance.

Karagiozova et al. [22] investigated the blast response of sandwich type panels with steel plates and polystyrene cores and compared them to panels with steel face plates and aluminium honeycomb cores. They found that as the impulse increased, the central displacement of both face skins increased as the honeycomb was able to transmit load through the cell structure to the back face. The permanent deflection of the back plate is influenced by the velocity attenuation properties of the core. Core efficiency, in terms of energy absorption, is an important factor for thicker cores. For panels of comparable mass, those with an aluminium honeycomb core perform better than those with polystyrene cores.

From Fig. 13, for the 13 mm sandwich panels with aluminium skins, the permanent front face displacement is 30% greater than the corresponding composite skinned system. Similar trends were observed in the thick sandwich structures. Figure 14 shows that the permanent front face displacement for the 25 mm thick sandwich structures (glass-fibre epoxy skinned) is lower than that of the 30 mm thick (aluminium skinned) sandwich structures. This behaviour is mainly governed by the properties of the face sheet. This suggests that the composite skins are superior at resisting blast loading than aluminium skinned sandwich structures.

The permanent back face displacement recorded in the 13 mm thick sandwich structures is shown in Fig. 15. It can be seen that the aluminium sandwich structures exhibits a greater displacement than the composite structures. For example, after an impulse of 2.7 N s, the back face permanent deformation of the centre panel aluminium skin is 6.91 mm whereas the composite skin is 4.91 mm. The permanent back face displacement recorded in the 25 and 30 mm thick sandwich structures is shown in Fig. 16. It can be seen that the aluminium sandwich structures exhibits slightly higher displacements than the composite structures.

Figures 17, 18, 19 and 20 show the effect of core thickness on the permanent front and back face displacements. It is clear that the graphs exhibit a similar appearance for the different thicknesses for both skin types. It was found that the thickness of the core does not affect the permanent displacement after blast loading.

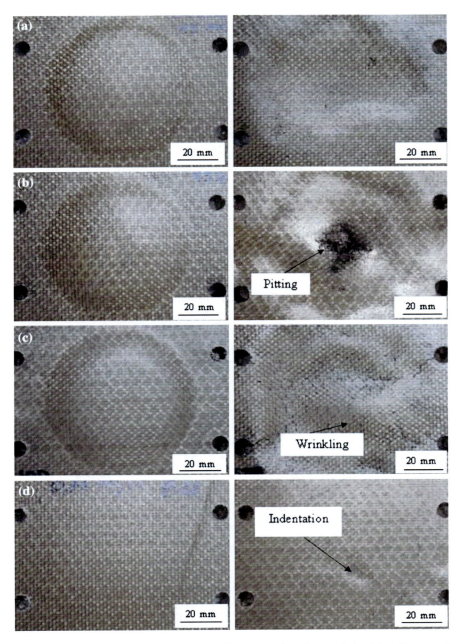

Fig. 11 Front and back surfaces of the glass-fibre epoxy face sheet and aluminium honeycomb core 25 mm thickness after blast testing (**a**) Impulse = 2.3 N s (Panel M1) (**b**) Impulse = 2.9 N s (Panel M5) (**c**) Impulse = 3.3 N s (Panel M6) (**d**) Impulse = 1.7 N s (Panel M2) (**e**) Impulse = 2.7 N s (Panel M3) (**f**) Impulse = 3.1 N s (Panel M4) (**g**) Impulse = 3.4 N s (Panel M8) (**h**) Impulse = 3.8 N s (Panel M7) (**i**) Impulse = 4.9 N s (Panel M9) (**j**) Impulse = 4.9 N s (panel M10)

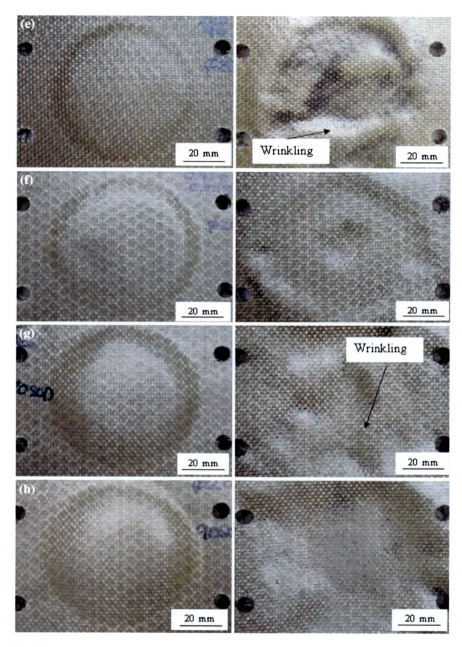

Fig. 11 (continued)

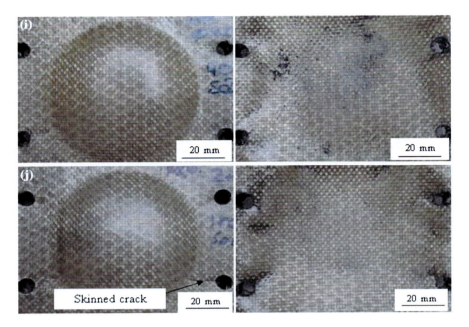

Fig. 11 (continued)

The effect of the core thickness on the permanent front face and back face displacement for the aluminium skinned sandwich structure is shown in Figs. 17 and 18. The graphs do not show any significant difference for the front face displacement, but the 13 mm core has a greater back face displacement than the thick sandwich structures. This result is in agreement with the observation that largest permanent deflections of the back plate occur for material with the highest density [19].

Figure 19 shows the permanent front face displacement for the glass-fibre/epoxy skinned sandwich structure. From the graph it is apparent that the 13 mm thick sandwich structure has a permanent displacement similar to that of the thick sandwich.

Figure 20 shows the permanent back displacement for the composite sandwich structures. It is evident that at lower impulses (between 2 and 3 N s) the displacement is very close for both thicknesses, but above an impulse of 3 N s, the 13 thick core exhibits a greater displacement than the 25 mm sandwich structure. For example, the composite skin in the thin sandwich structures (13 mm thick core) exhibited a back face deflection of 13.71 mm following an impulse of 4.7 N s, whereas that specimen of the thick sandwich structures was 10.12 mm following an impulse of 4.9 N s. Clearly, the thicker core has a high rigidity in flexure resulting in a lower deflection under blast loading. This is consistent with previous findings on the impact response of sandwich structure by Park et al. [23] and

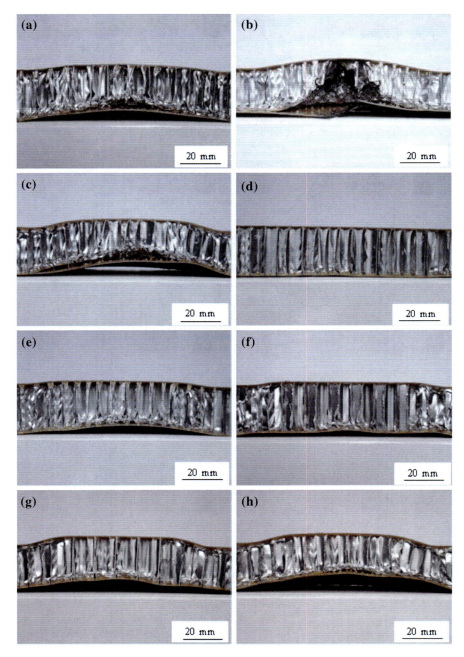

Fig. 12 Cross-section views of the glass-fibre epoxy/aluminium honeycomb core sandwich (core thickness = 25 mm) (**a**) Impulse = 2.3 N s (**b**) Impulse = 2.9 N s (**c**) Impulse = 3.3 N s (**d**) Impulse = 1.7 N s (**e**) Impulse = 2.7 N s (**f**) Impulse = 3.1 N s (**g**) Impulse = 3.4 N s (**h**) Impulse = 3.8 N s (**i**) Impulse = 4.9 N s (**j**) Impulse = 4.9 N s

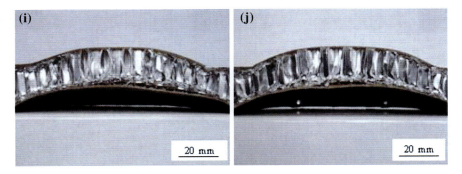

Fig. 12 (continued)

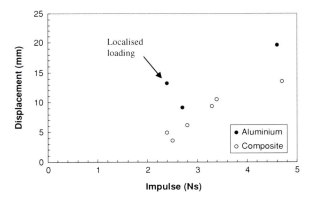

Fig. 13 Permanent front face displacement versus impulse for the 13 mm thick sandwich structures

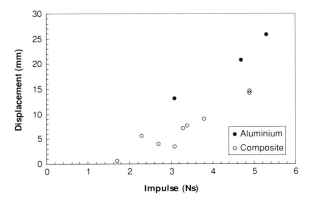

Fig. 14 Permanent front face displacement versus impulse for the 25 mm thick (glass-fibre epoxy skinned) and 30 mm thick (aluminium skinned) sandwich structures

Dear et al. [24]. They found that the impact resistance of a sandwich structure is greatly influenced by the face sheet type and core thickness. Similar observations were reported by Hazizan and Cantwell [25], who conducted drop-weight impact test on glass-fibre reinforced epoxy skin/aluminium core sandwich structure and

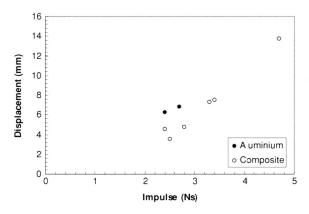

Fig. 15 Permanent back face displacement versus impulse for the 13 mm thick sandwich structures

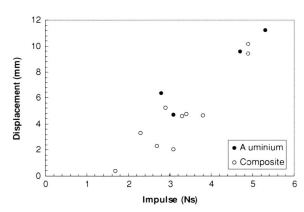

Fig. 16 Permanent back face displacement versus impulse for the 30 mm thick (*aluminium*) and 25 mm thick (*composite*)

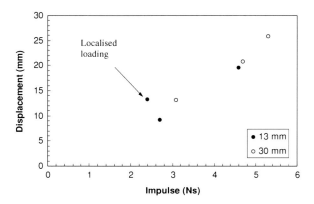

Fig. 17 Permanent front face displacement versus impulse for the aluminium skinned sandwich structures(*Localised loading*)

measured the maximum impact force for 13 and 25 mm thick aluminium sandwich structures. For example, at a 200 mm span, the effect of changing the core thickness was significant, with the maximum impact force increasing from 354 N

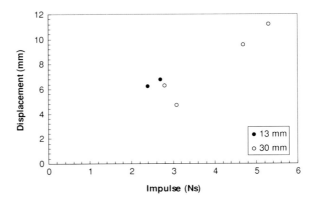

Fig. 18 Permanent back face displacement versus impulse for the aluminium skinned sandwich structures

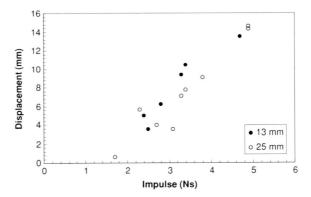

Fig. 19 Permanent front face displacement versus impulse for the glass-fibre/epoxy skinned sandwich structures

to 449 N as the core thickness was increased from 13 to 25 mm. Again, this fact is due to the higher stiffness of the structure with a thicker core, since the bending stiffness, D, of the sandwich beam increases as the thickness increases. Radford et al. [26] investigated the dynamic responses of clamped circular monolithic and sandwich plates of equal areal mass by loading the plates at their mid-span with metal foam projectiles. The sandwich plates were based on stainless steel face sheets with an aluminium alloy metal foam cores. It is found that the sandwich plates offer a higher shock resistance than monolithic plates of equal mass. Further, the shock resistance of the sandwich plates increases with increasing thickness of sandwich core.

The permanent deformation of the front and back faces was greater in the aluminium skin than in the composite skinned sandwich structures, a reflection of the ability of the aluminium to undergo plastic deformation. The same results were reported previously by Dear et al. [24] following impact tests on woven glass-fibre impregnated with epoxy resin and a lightweight aluminium honeycomb core and aluminium alloy sheet bonded to lightweight aluminium honeycomb core.

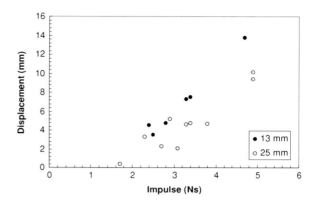

Fig. 20 Permanent back face displacement versus impulse for the glass-fibre/epoxy skinned sandwich structures

This finding is supported by Shin et al. [27], who investigated on the low-velocity impact response of four different types of sandwich structures. Impact parameters such as the maximum contact force, contact time, deflection at the peak load and absorbed energy were evaluated and compared for different types of sandwich panels. The impact test results showed that sandwich panels with woven glass fabric/epoxy face sheet offer a superior impact damage resistance than sandwich panels with metal aluminium face sheets.

Zhu et al. [28] investigated the effect face-sheet thickness, cell size and foil thickness of the honeycomb and mass of charge on the structural response of sandwich panels loaded by blast of face sheet and core configuration. Based on a quantitative analysis, it has also been found that the face sheet thickness and relative density of core structure can significantly influence the back face deformation. By adopting thicker skins and a honeycomb core with a higher relative density, the deflection of the back face can be reduced. Also, for a given panel configuration, it is evident that the back face deflection increases with impulse, is an approximately linear fashion.

The blast performance of the sandwich structures was compared to that of the plain composite laminates by dividing the impulse to give complete failure of the sandwich structure by its areal density. However, in spite of the fact that all four types of sandwich structure had been tested to impulses close to that required to completely destroy the laminates, none of them actually failed completely. For this reason, the calculated values of specific impulses represent lower bounds for the data. Figure 21 compares the specific impulses for the sandwich structures with those for the plain composites discussed earlier. The arrows in the figure indicate that the exact values of specific impulse for failure are higher. In spite of this, it is evident that the performance of the sandwich panels is similar to that of the plain composites, suggesting that, in terms of specific impulse to give complete failure, there is no significant benefit in employing sandwich structures. Indeed, there is the disadvantage that the sandwich structures are thicker than the plain laminates.

Fig. 21 The impulse to initiate lower surface fracture in the sandwich structures and composite panels normalised by areal density

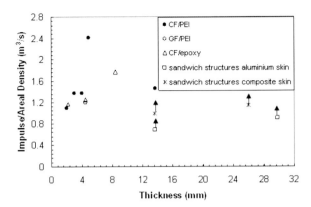

The energy dissipated in crushing the core of the sandwich structures was estimated using the information available in the load–displacement traces shown in Figs. 2 and 3. Here, the energy dissipated during crush was determined from the area under the load–displacement trace. This was then normalised by the crush depth and then by the planar area of the test sample (approximately 900 mm^2 in the present case). This gave an energy per unit volume of 2.04 x10^6 J/m^3 for the composite sandwich structure and 2.5 x 10^6 J/m^3 for the aluminium-skinned system. These values were then used to estimate the energy used in crushing the honeycomb during the blast test. Here, the average crush was determined from the cross-sections and this was multiplied by the planar area that was crushed. This gave a value, in Joules, for the energy dissipated in crushing the core material. Although only an approximate approach, this technique does give an indication of the energy absorbed in this mechanism.

The estimated energy absorbed in crushing the aluminium honeycomb in the sandwich structures is given in Table 4. Examination of the table indicated that energies up to 200 J have been absorbed in this failure mechanism. The information in this table is plotted in graphical form in Figs. 22, 23, 24 and 25 where the energy absorbed is plotted against impulse to investigate the effect of varying core thickness and skin material. Figures 22 and 23 show the effect of core thickness on energy absorption for the aluminium and composite-skinned sandwich structures respectively. As expected, the energy absorbed in crushing increases with impulse. Extrapolating the data back to zero energy suggests that the critical impulse for initiating energy absorption in this mechanism is around 2 N s. From Figs. 22 and 23, it is evident that the thicker sandwiches absorb more energy than their thinner counterparts. This would be expected since there is a greater volume of core material in the 30 mm thick laminates. Figures 24 and 25 show the effect of skin type on the energy absorbing process. Clearly, there are more data points for the composite systems than for the aluminium-skinned laminates. In spite of this, the evidence suggests that the sandwiches with aluminium skins absorb slightly more energy in crushing of the core than their

Table 4 Summary of the estimated energies absorbed in crushing the cores of the sandwich structures

	Specimen	SOD (mm)	Impulse (Ns)	Energy (J)
Aluminium 13 mm	J1	45	2.4	21.2
	J2	90	2.7	32.9
	J3	180	4.6	63.1
Aluminium 30 mm	K1	45	2.8	54.7
	K2	90	3.1	97.9
	K3	90	4.7	165.1
	K4	180	5.3	201.7
Composite 13 mm	L1	45	2.4	20.8
	L2	90	2.8	15.4
	L3	180	2.5	7.8
	L4	90	3.3	32.6
	L5	180	3.4	37.4
	L6	180	4.7	56.6
Composite 25 mm	M1	45	2.3	22.9
	M2	90	1.7	1.8
	M3	90	2.7	13.9
	M4	180	3.1	7.5
	M5	45	2.9	46.4
	M6	65	3.3	78.5
	M7	90	3.8	56.1
	M8	180	3.4	29.8
	M9	90	4.9	122.2
	M10	180	4.9	99.8

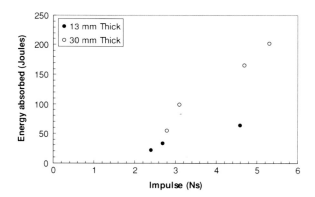

Fig. 22 Energy absorbed by the core versus impulse for the aluminium-skinned sandwich structures

composite counterparts. This may be associated with the ability of the uppermost aluminium skin to undergo greater plastic deformation during blast than the composite skin. It is likely that the uppermost composite skin would fail at relatively low strains due to the limited strain capability of the glass-fibres. In contrast, the aluminium alloy can deform to higher strains and therefore crush the core to a greater degree before failing. Given the crudeness of the approach, the trends in the data appear reasonably clear, giving a useful first estimate of the energy absorbed in this process.

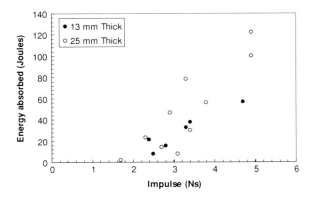

Fig. 23 Energy absorbed by the core versus impulse for the composite-skinned sandwich structures

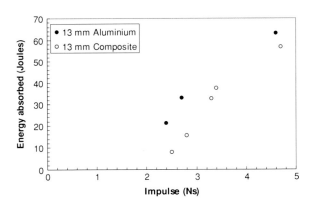

Fig. 24 Energy absorbed by the core versus impulse for the sandwich structures with a 13 mm thick core

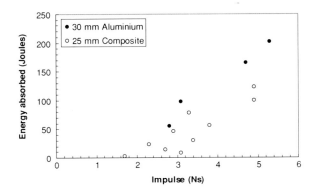

Fig. 25 Energy absorbed by the core versus impulse for the sandwich structures with a 30 mm thick core

8 Conclusion

Blast tests on aluminium honeycomb sandwich structures have shown that the structures with glass-fibre reinforced epoxy skins offer a superior blast resistance than those with aluminium alloy skins. An examination of the damaged sandwich structures highlighted failure mechanisms including core crushing, fibre fracture, pitting, rear surface tearing on the back surface, and top surface indentation. Off these, it appears that core crushing is the most important energy-absorbing mechanism. Tests on sandwich structures with different core thickness have shown that sandwich structures with thicker cores offer a superior energy-absorbing capacity under blast loading.

Acknowledgments The authors are grateful to the Malaysian Government EPSRC (Grant EP/E037887/1) for funding part of this work.

References

1. Zhu, F., Zhao, L., Lu, C., Wang, Z.: Deformation and failure of blast-loaded metallic sandwich panel – experimental investigations. Accepted for publication in International Journal of Impact Engineering
2. Hanssen, A.G., Enstock, L., Langseth, M.: Close range blast loading of aluminium foam panels. International J. Imp. Eng. **27**, 593–618 (2002)
3. Karagiozova, D., Nurick, G.N., Langdon, G.S., Yuen, S.C.K., Chi, Y., Bartle, S.: Response of flexible sandwich-type panels to blast loading. Accepted for publication in Composites Science and Technology
4. McKown, S., Shen, Y., Brookes, W.K., Sutcliffe, C.J., Cantwell, W.J., Langdon, G.S., Nurick, G.N., Theobald, M.D.: The quasi-static and blast loading response of lattice structures. Int. J. Imp. Eng. **35**, 795–810 (2008)
5. Radford, D.D., McShane, G.J., Deshpande, V.S., Fleck, N.A.: The response of clamped sandwich plates with metallic foam cores to simulated blast loading. Int. J. Sol. Struct. **43**, 2243–2259 (2006)
6. Fleck, N.A., Deshpande, V.S.: The resistance of clamped sandwich beams to shock loading. J. Appl. Mech. ASME **71**, 386–401 (2004)
7. Xue, Z., Hutchinson, J.W.: A comparative study of blast-resistant metal sandwich plates. Int.J. Imp. Eng. **30**, 1283–1305 (2004)
8. Sriram, R., Vaidya, U.K.: Blast impact response of aluminium foam sandwich composites. J. Mater. Sci. **41**, 4023–4039 (2006)
9. Hazizan, M.A., Cantwell, W.J.: The low velocity impact response of an aluminium honeycomb sandwich struct. Compos.: Part B **34**, 679–687 (2003)
10. Chi, Y.C.: The response of honeycomb sandwich panels to blast load. M.Sc. dissitation, University of Cape Town (2008)
11. Aminanda, Y., Castanie, B., Barrau, J.-J., Thevenet, P.: Experimental analysis and modelling of the crushing of honeycomb cores. Appl. Compos. Mater. **12**, 213–227 (2005)
12. Othman, A.R., Barton, D.C.: Failure initiation and propagation characteristics of honeycomb sandwich composites. Compos. Struct. **85**, 126–138 (2008)
13. Wu, E., Jiang, W.S.: Axial crush of metallic honeycomb. Int. J. Imp. Eng. **9**, 439–456 (1997)
14. Paik, J.K., Thayamballi, A.K., Kim, G.S.: The strength characteristics of aluminium honeycomb sandwich panels. Thin-Wall. Struct. **35**, 205–231 (1999)

15. Langdon, G.S., Cantwell, W.J., Nurick, G.N.: The blast response of novel thermoplastic-based fibre-metal laminates–some preliminary results and observations. Compos. Sci. Technol. **65**, 861–872 (2005)
16. Teeling-Smith, R.G., Nurick, G.N.: The deformation and tearing of circular plates subjected to impulsive loads. Int. J. Imp. Eng. **11**, 77–92 (1991)
17. Nurick, G.N., Martin, J.B.: Deformation of thin plates subjected to impulsive loading—A review; Experimental studies. Int. J. Imp. Eng. **8**, 170–186 (1989)
18. Xue, Z., Hutchinson, J.W.: Crush dynamics of square honeycomb sandwich cores. Int. J. Num. Method. Eng. **65**, 2221–2245 (2005)
19. Dharmasena, K.P., Wadley, H.N.G., Xue, Z., Hutcinson, J.W.: Mechanical response of metallic honeycomb sandwich panel structures to high-intensity dynamic loading. Accepted for publication in Int. J. Imp. Eng
20. Yahya, M.Y., Cantwell, W.J., Nurick, G.N., Langdon, G.S.: The blast behavior of fibre reinforced thermoplastic laminates, pp. 2275–2297. J. Compos. Mater. SAGE Publications (2008)
21. Yahya, M.Y., Cantwell, W.J., Nurick, G.N., Langdon, G.S.: The blast resistance of a woven carbon fibre reinforced epoxy composite. J. Compos. Mater. SAGE Publications. 45 (7), 789–801 April (2011)
22. Karagiozova, D., Nurick, G.N., Langdon, G.S. , Yuen, S.C.K., Chi, Y., Bartle, S.: Response of flexible sandwich-type panels to blast loading. Accepted for publication in Composites Science and Technology
23. Park, J.H., Ha, S.K., Kang, K.W., Kim, C.W., Kim, H.S.: Impact damage resistance of sandwich structure subjected to low velocity impact. J. Mater. Process. Technol. **201**, 425–430 (2008)
24. Dear, J.P., Lee, H., Brown, S.A.: Impact damage processes in composite sheet and sandwich honeycomb materials. Int. J. Imp. Eng. **32**, 130–154 (2005)
25. Hazizan, M.A., Cantwell, W.J.: The low velocity impact response of an aluminium honeycomb sandwich structure. Composites: Part B **34**, 679–687 (2003)
26. Radford, D.D., McShane, G.J., Deshpande, V.S., Fleck, N.A.: The response of clamped sandwich plates with metallic foam cores to simulated blast loading. International J. Sol. Struct. **43**, 2243–2259 (2006)
27. Shin, K.B., Lee, J.Y., Cho, S.H.: An experimental study of low-velocity impact responses of sandwich panels for Korean low floor bus. Compos. Struct. **84**, 228–240 (2008)
28. Zhu, F., Zhao, L., Lu, G., Wang, Z.: Deformation and failure of blast-loaded metallic sandwich panels—Experimental investigations. Accepted for publication in International Journal of Impact Engineering

The High Velocity Impact Response of Self-Reinforced Polypropylene Fibre Metal Laminates

M. R. Abdullah and W. J. Cantwell

Abstract The high velocity impact response of a range of polypropylene-based fibre-metal laminate (FML) structures has been investigated. Initial tests were conducted on simple FML sandwich structures based on 2024-O and 2024-T3 aluminium alloy skins and a Self-Reinforced Polypropylene (SRPP) composite core. Here, it was shown that laminates based on the stronger 2024-T3 alloy offered a superior perforation resistance to those based on the 2024-O system. Tests were also conducted on multi-layered materials in which the composite plies were dispersed between more than two aluminium sheets. For a given target thickness, the multi-layered laminates offered a superior perforation resistance to the sandwich laminates. The perforation resistances of the various laminates investigated here were compared by determining the specific perforation energy (s.p.e) of each system. Here, the sandwich FMLs based on the low density SRPP core out-performed the multi-layer systems, offering s.p.e.'s roughly double that exhibited by a similar Kevlar-based laminate. A closer examination of the panels highlighted a number of failure mechanisms such as ductile tearing, delamination and fibre failure in the composite plies as well as permanent plastic deformation, thinning and shear fracture in the metal layers. Finally, the perforation threshold of all of the FML structures was predicted using the Reid–Wen perforation model. Here, it was found that the predictions offered by this simple model were in good agreement with the experimental data.

M. R. Abdullah (✉)
Department of Solid Mechanics and Design,
Faculty of Mechanical Engineering, Universiti Teknologi Malaysia,
81310 Johor, Bahru, Malaysia
e-mail: ruslan@fkm.utm.my

W. J. Cantwell
Department of Engineering, University of Liverpool,
Liverpool L69 3GH, UK

Keywords Fibre–metal laminates · Impact behaviour · Failure mechanism

1 Introduction

Lightweight composite materials are currently finding extensive use in a wide range of load-bearing engineering applications. Previous work has shown that fibre-reinforced polymers such as carbon fibre reinforced plastic (CFRP) offer outstanding strength and stiffness properties as well as a superior resistance to long–term fatigue cycling [1–3]. One of the frequently cited limitations of fibre-reinforced plastics is their relatively poor resistance to localised impact loading [4–6]. Indeed, low velocity impact tests have shown that CFRP suffers a significant reduction in its load-bearing properties following impact at energies as low as 2 Joules [7]. In addition, a number of workers have shown that many composites offer a relatively poor resistance to penetration and perforation, failure processes that are frequently encountered during high velocity impact loading by a low mass projectile [8–10]. In an attempt to overcome these limitations, a number of researchers have investigated the possibility of developing hybrid structures based on, for example, mixtures of glass, Kevlar and carbon fibres [9, 10]. Another interesting possibility is to develop metal-composite systems such as fibre-metal laminates (FMLs) based on thin layers of metal and fibre-reinforced composite material. At present, systems such as glass fibre/aluminium (GLARE), aramid fibre/aluminium (ARALL) and carbon fibre/aluminium (CALL) are attracting interest from a wide range of engineering sectors. The mechanical properties of epoxy-based fibre-metal laminates have been investigated in a number of studies [11–13].

Alderliesten [12] conducted a series of fatigue tests on GLARE and plain aluminium alloy specimens and found that crack growth rates in fibre-metal systems were between one and two orders of magnitude lower than in aluminium alloy samples. Vlot et al. [14] conducted impact tests on GLARE, a monolithic aluminium alloy and carbon fibre reinforced PEEK. Their results indicated that the damage threshold energy for this multi-layered material was significantly greater than values offered by more traditional engineering materials.

The first generation of thermosetting-based fibre-metal laminates suffer a number of key limitations including long processing cycles, low interlaminar fracture toughness properties as well as difficulties associated with repair. In an attempt to overcome many of these problems, a number of novel FMLs based on thermoplastic matrices have been developed and tested [13, 15]. Thermoplastic-based fibre-metal laminates offer a number of advantages including very short processing times, ease of forming, improved chemical resistance, excellent repairability and superior interlaminar fracture toughness properties. Extensive testing on a glass fibre-reinforced polypropylene FML has shown that this system offers an excellent resistance to low and high velocity impact loading conditions [15]. A subsequent study suggested that FMLs based on a polypropylene fibre reinforced polypropylene matrix offer an excellent resistance to localised high velocity impact loading [16]. In addition, self-reinforced polypropylene systems combine the versatility and recyclability of thermoplastics with the superior mechanical properties offered by fibre-reinforced composites.

The aim of this research project is to investigate the impact resistance of a lightweight fibre-metal laminate based on thin layers of an aluminium alloy and a self-reinforced polypropylene composite. Particular attention centres on investigating the influence of varying the stacking sequence on the perforation resistance of these hybrid laminates.

2 Experimental Procedure

The fibre-metal laminates investigated in this study were based on a SRPP composite (Curv from Propex Fabric) and two types of aluminium alloy (a 0.6 mm thick 2024-O alloy and a 0.8 mm 2024-T3 alloy). Before laminating, the aluminium and the SRPP sheets were cleaned with acetone in order to remove any dust, oil or grease. Optimum adhesion across the composite metal interface was ensured by placing a 60 micron thick thermoplastic interlayer (Gluco from Gluco Ltd, Leeds, UK) at each common interface. This low-viscosity polymer interlayer offers excellent flow characteristic and improved interlocking with the surface of the metal.

Table 1 gives a summary of the materials investigated in this research study. The fibre-metal laminates were manufactured by stacking the appropriate number of composite plies, aluminium sheets and interlayer films in a picture-frame mould and placing the stack in a press. The laminates were then heated to 165°C at a rate of approximately 10°C/min before being cooled to room temperature at a rate of approximately 5°C/min. Once the press had cooled to a temperature below 60°C, the panel was removed from the mould and visually inspected for defects. A schematic illustration of a 4/3 FML is shown in Fig. 1. Details of the lay-ups and stacking configurations studied in this research programme are given in Table 2

A nitrogen gas gun was used to evaluate the high velocity impact response of the FMLs. Figure 2 shows a schematic representation of the gas gun test facility used in this study. Square plates with an edge length of 100 mm were clamped in a steel support with a 75 mm square aperture. The rig was then bolted to a steel block and a velocimeter was placed between the end of the barrel and the target to measure the velocity of the projectile (Fig. 3a, b). Here, the velocimeter ensured a minimum accuracy in measuring the impact velocity of 99.6%. Impact testing was conducted using a 46.7 g steel projectile with a 12.7 mm diameter hemispherical head (Fig. 3c). The velocity of the projectile was controlled by adjusting the pressure in the main chamber of the gas gun shown in Fig. 2.

Impact testing was conducted over a range of impact energies until complete perforation of the FML target was achieved. The resulting perforation energy was then used to calculate the specific perforation energy of the target by normalising the measured perforation energy by the areal density of the target.

Table 1 Summary of the materials investigated in this research programme

Material	Grade	Supplier	Density (kg/m³)
Aluminium Alloy	2024-T3 (0.8 mm thick)	Alcan	2,770
	2024-O (0.6 mm thick)	Alcan	2,770
Curv	Woven polypropylene fibre/Polypropylene matrix	Propex Fabric	920
Gluco	Polypropylene adhesive film 55 g/m²	Gluco Ltd.	

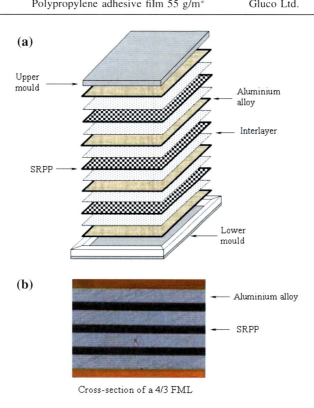

Fig. 1 Schematic illustration of the manufacturing procedure for a 4/3, Self-reinforced polypropylene fibre-metal laminate

In order to achieve a greater understanding of the impact behaviour of the FMLs, a series of impact tests were carried out on the plain aluminium alloy and the SRPP composite. After testing, the specimens were removed from the gas gun, sectioned, polished and then viewed under an optical microscope in order to elucidate the failure mechanisms during impact. The specimens were sectioned through the point of impact using a band-saw and ground using 1200 grit silicon carbide paper.

Following impact testing, the maximum out-of-plane displacement in the impacted samples was measured in order to characterise the impact response of

Table 2 Details of the lay-ups and stacking configurations investigated in this study

Lay-up	Stacking configuration	Average thickness (mm)	Areal density (kg/m^2)
Al 2024-O and SRPP composite			
O2/1	Al/SRPP/Al	1.75	4.06
O3/2	Al/SRPP/Al/SRPP/Al	2.85	6.28
O4/3	Al/SRPP/Al/SRPP/Al/SRPP/Al	3.88	8.48
O5/4	Al/SRPP/Al/SRPP/Al/SRPP/Al/SRPP/Al	4.89	10.69
Al 2024-T3 and SRPP composite			
T32/1	Al/SRPP/Al	2.06	4.97
T33/2	Al/SRPP/Al/SRPP/Al	3.28	7.62
T34/3	Al/SRPP/Al/SRPP/Al/SRPP/Al	4.53	10.31
T35/4	Al/SRPP/Al/SRPP/Al/SRPP/Al/SRPP/Al	5.73	12.96

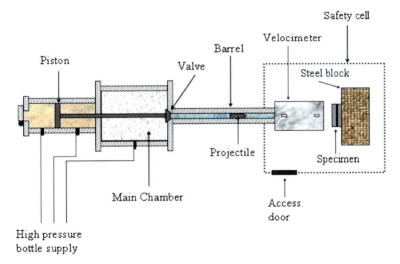

Fig. 2 Schematic illustration of the gas gun used for conducting the high velocity impact tests

the FML specimens. Here, tangents were drawn to the deformed portions of the specimen, and the displacement, δ, was measured from the point of intersection of these tangents to the initial, undeformed shape, as shown in Fig. 4. Although this method is somewhat crude, it does yield useful information on the energy-absorbing capacity of the specimens.

3 The Perforation Model

The perforation threshold energy for the FMLs was predicted using the perforation model developed by Reid and Wen [17]. Here, it is assumed that when a laminate is perforated by a projectile at high velocities, the mean pressure (σ) applied to the projectile by the target can be separated into two components: one associated with

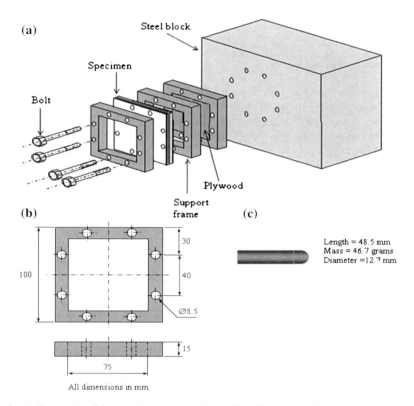

Fig. 3 (a) Schematic of the specimen support (b) details of the support frame shown in plan and side view and (c) the hemispherical-ended projectile used during impact testing

Fig. 4 Method employed for measuring the permanent displacement of the target

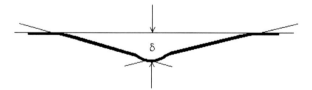

the cohesive static resistive pressure due to elastic–plastic deformations within the laminate (σ_s), and a second associated with the dynamic resistive pressure (σ_d) associated with loading rate effects [17]. This can be written as:

$$\sigma = \sigma_s + \sigma_d \quad (1)$$

It is also assumed that the cohesive static resistive pressure is equal to the static linear elastic compression limit (σ_e) in the through-thickness direction, i.e. $\sigma_s = \sigma_e$, and that the dynamic resistive pressure (σ_d) can be written as:

$$\sigma_d = \Gamma(\rho_\tau/\sigma_e)^{1/2} V_i \sigma_e \quad (2)$$

The High Velocity Impact Response   225

Where, ρ_t and V_i represent the density of the laminate and the initial projectile velocity respectively. The term Γ is a constant having a value of 1.5 for a hemispherical projectile [17].

Equation 1 can therefore be written in the form:

$$\sigma = \left[1 + \Gamma\sqrt{\frac{\rho_t}{\sigma_e}}V_i\right]\sigma_e \qquad (3)$$

Reid and Wen showed that the perforation energy of a composite laminate when struck transversely by a rigid projectile at high velocity is given as [17]:

$$E_p = \frac{\pi D^2 T \sigma}{4} \qquad (4)$$

where T is the thickness of the laminate and D is the diameter of the projectile.

Substituting Eq. 3 into Eq. 4 gives:

$$E_p = \frac{\pi D^2 T}{4}\left[1 + \Gamma\sqrt{\frac{\rho_t}{\sigma_e}}V_b\right]\sigma_e \qquad (5)$$

Knowing that $E_p = \frac{1}{2}mV^2$, the perforation velocity can be written as [17]:

$$V_b = \frac{\pi\Gamma\sqrt{\rho_t\sigma_e}D^2T}{4m}\left[1 + \sqrt{1 + \frac{8m}{\pi\Gamma^2\rho_t D^2 T}}\right] \qquad (6)$$

This equation was used to predict the perforation threshold of the FMLs examined in this study.

4 Results and Discussion

4.1 Characterisation of the Residual Deformation in the Impacted FMLs

The impact response of the FMLs was investigated by measuring the residual displacement in the impact-loaded structures. Generally, the residual deformation of an FML specimen following impact testing was similar to that of plain aluminium alloy. Clearly, due to the ductility of the aluminium plies, deformation of the FMLs was not limited to the zone surrounding the impact location. It is interesting to note that only that part of the specimen clamped within the test frame was undeformed at the end of impact event, whereas the remainder of the sample exhibited a pronounced concavity. In contrast, examination of the plain SRPP specimens showed that residual displacements in these laminates were very small up to the perforation threshold. It is likely that the composite laminates exhibit a very localised form of target response following high velocity impact loading. These

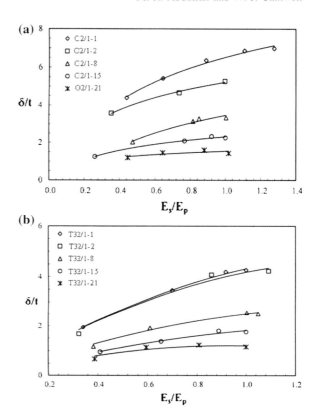

Fig. 5 Non-dimensional maximum permanent displacement versus normalised impact energy of the FMLs based-SRPP (a) O2/1 and (b) T32/1

findings have been reported elsewhere where it has been shown that composites revert back substantially to their original shape after impact, with only a small region of permanent damage occurring around the entrance and exit points [18]. This suggests that elastic deformations in the SRPP plies were responsible for absorbing energy during the impact tests. In this study, no attempt was made to measure the permanent displacement of the plain composite.

Figure 5 summarises the non-dimensional permanent deformation recorded in the 2/1 FMLs following high velocity impact testing. Here, the residual displacement normalised by the initial specimen thickness is plotted against the normalised impact energy (the incident energy, E_s divided by the perforation energy, E_p). This allows a comparison of a wide range of stacking sequences and laminate thicknesses. The figure shows that the residual permanent displacement in the targets increases with increasing incident impact energy, reaching a plateau as the perforation threshold is approached. At low and intermediate incident energies, the permanent displacement in thin and intermediate thickness laminates increases rapidly before tending to plateau. Apparently, at normalised energies of approximately 1.0, there is no significant increase in normalised displacement. These roughly constant values of normalised displacement are due to the fact that

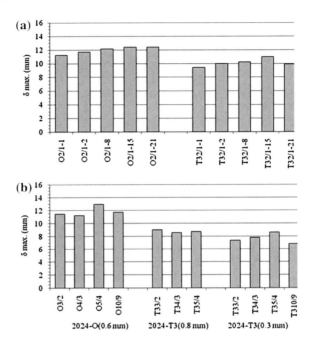

Fig. 6 Maximum permanent out-of-plane displacement at the perforation threshold for the 2024-O and 2024-T3 laminates

the absorbed energy remains constant beyond the perforation velocity, i.e. the maximum energy which the laminates can absorb during the impact event occurs at the perforation threshold [19, 20]. Above this limit, the absorbed energy remained constant with increasing impact velocity. In other words, when the impact velocity is greater than the perforation velocity, the energy absorbed by the laminates is the same or possibly less than at the perforation energy. Furthermore, under low impact velocity, the contact time between the projectile and the laminate is relatively long and the incident momentum high. As a result, the whole laminate has time to respond and there is sufficient time for the laminate to undergo progressive deformation. In contrast, faster projectiles penetrate the laminate in a shorter time and it is likely that the particle may not have enough time to accelerate the whole structure and therefore most of the available energy is dissipated over a small zone immediate to the point of contact. In the case of thicker laminates, the permanent displacement increases slowly before tending to a constant value as the perforation threshold is approached. Clearly, the thick laminates offer a very high bending stiffness leads to much higher impact forces thus causing energy absorption through damage in the laminate.

Figure 6 shows the maximum permanent displacement at the perforation threshold for the 2024-T3 and 2024-O laminates. Again, the critical out-of-plane deformation, δ_{max}, for the 2024-O laminates is roughly constant for all of the laminates, averaging approximately 12 mm regardless of the plate configuration or target thickness. Similarly, the average of δ_{max} for the 2024-T3 system is also roughly constant, averaging approximately 10 mm over the range of laminates investigated. It is believed that the lower value of δ_{max} for the T3 FMLs results from the increased thickness of these layers, the lower strain to failure of this alloy as well as the change in failure mode (thinning to shear)during impact. The fact that the maximum rear surface deformation of the targets remains roughly constant as the thickness of the target is increased suggests that the aluminium layers may deform independently of each other during the perforation process. Here, the low modulus of the self-reinforced composite allows the aluminium plies to deform in a quasi-independent manner when loaded in flexure. As observed previously, the out of plane displacement of these plies can lead to localised membrane stretching which in turn leads to enhanced energy absorption. This was investigated further by conducting tests on aluminium samples that were bonded using an unreinforced polypropylene adhesive. The perforation energies of these systems were not as impressive as their composite counterparts indicating that the composite makes a positive contribution to the perforation process rather than simply acting as a low modulus adhesive holding the aluminium sheets together. Indeed, the room temperature Izod fracture energy of the composite is very high (4750 J/m)[21] suggesting that considerable energy is also absorbed in fracturing these layers during the perforation process. This will be investigated further in the following section where the volume fraction of composite in the FMLs is varied.

4.2 Impact Failure Mechanisms

The failure processes in the impact-loaded FMLs were initially investigated by examining the front and rear surfaces of the impact-damaged samples. Initial fracture in the 2024-O 2/1 FML following a 33.6 J impact is shown in the low magnification micrographs presented in Fig. 7a. Here, damage takes the form of a large top surface dent and a localised crack parallel to the rolling direction of the rearmost aluminium layer. Similar failure mechanisms have been observed in GLARE laminates tested under low velocity impact conditions [22]. With increasing energy, the length of the rear surface crack increased as did the size of the top surface dent. At the perforation threshold, the passage of the projectile through the target produced a relatively clean hole with a diameter similar to that of the steel projectile, Fig. 7b. Clearly, the perforation process involves significant local ductility in both the aluminium and composite plies as well as fracture of the constituent materials. The mechanisms of damage initiation in the thicker 2024-O 4/3 laminate were similar to those observed in its thinner 2/1 counterpart with initial failure occurring in the rear surface layer parallel to the rolling direction of the aluminium, Fig. 8a. At higher energies, a large dent and then a circular crack develop around the impactor and the uppermost composite and metal plies begin to be pushed through the rear surface opening, Fig. 8b. Finally, at the perforation threshold, the projectile passes through the target leaving a relatively clean hole and limited petalling at the rear surface (Fig. 8c). Figure 9a shows damage in the

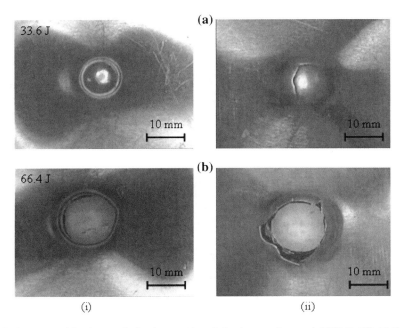

Fig. 7 Low magnification optical micrographs of the impact-damaged 2024-O 2/1 FMLs (**i**) impacted surface and (**ii**) rear surface

5/4 T3 laminate where initial failure again took the form of top surface denting coupled with a split in the lowermost aluminium ply parallel to the rolling direction. At the perforation threshold, a circular hole was visible in the top surface ply and localised petalling of the rear aluminium layer. Typically, there were three dominant petals in these perforated samples with minor ones propagating between them. In the 4/3 and 5/4 laminates, the projectile frequently remained embedded in the target at impact energies just below the perforation threshold. It is worth noting that in the 4/3 and 5/4 laminates, the projectile frequently remained embedded in the target at impact energies just below the perforation threshold. A careful examination of the impact chamber indicated that the penetration and perforation processes produced very little debris. In most cases, only a small number of composite fragments and a dish-shaped plug were observed, Fig. 10.

The failure processes in the impacted specimens were further investigated by sectioning a number of samples transversely through the center of impact after testing. The cross-sections were then polished and examined under an optical microscope. The subsequent development of damage with increasing impact energy in the 2/1 and 5/4 FML configurations is presented in Figs. 11 and 12 respectively. In the 2024-O 2/1 laminate at lower impact energies, the lower surface aluminium fractures and the upper surface ply exhibits localised thinning around the point of impact, Fig. 11a. It is believed that this thinning process results from membrane stretching and subsequent yielding in the thin aluminium plies

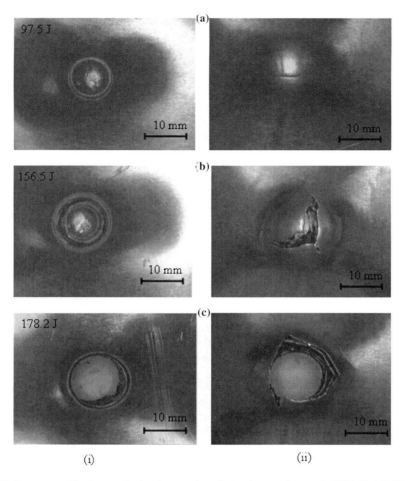

Fig. 8 Low magnification optical micrographs of the impact-damaged 2024-O 4/3 FMLs (i) impacted surface and (ii) rear surface

during the impact process. Following an impact energy of 46.2 J, Fig. 11b, fracture of upper and lower aluminium plies is observed as well as localised fracture in composite core of this sandwich structure. Finally, at the perforation threshold, projectile passes through the target, a process that involves significant plastic deformation in the zone of material immediate to the impact location Fig. 11c. Similar failure processes were observed in the corresponding 2024-T3 laminates, Figs. 11d–f although evidence of shear fracture in the aluminium layers was observed at intermediate energies. The processes of damage initiation and subsequent target perforation were similar in the 3/2 and 4/3 laminates. Once again thinning of the aluminium plies was observed around the point of impact in the

The High Velocity Impact Response

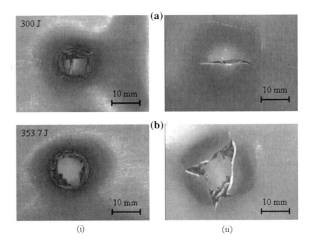

Fig. 9 Low magnification optical micrographs of the impact-damaged 2024-T3 5/4 FMLs (**i**) impacted surface and (**ii**) rear surface

Fig. 10 Dish-shaped plug ejected from an impacted-loaded target

2024-O laminates at intermediate impact energies whereas shear fracture was more pronounced in the T3 systems. Typical examples of localised thinning and shear fracture of the aluminium plies are shown in Fig. 13. A closer inspection showed that delamination and ductile tearing within the composite plies were more pronounced in the T3 systems than in their 2024-O counterparts. Interestingly, Hagenbeek stated that transverse shear failure such as that observed in the T3 system can act as a precursor to delamination in FMLs [22]. Failure in the 5/4 laminates tended to initiate in the upper part of the target immediate to the point of impact, Fig. 12. Here, the passage of the projectile through the target precipitated failure of the aluminium alloy immediately under the point of impact. As previously observed, localised thinning of the aluminium was observed in the 2024-O FMLs whereas transverse shear fracture was apparent in the T3 laminates.

Fig. 11 Low magnification optical micrographs of polished sections of 2/1 FMLs subjected to various impact energies (**i**) 2024-O (**ii**) 2024-T3

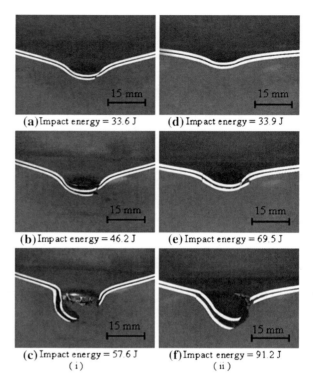

Once again, delamination and ductile tearing within the composite layers was more pronounced in the T3 system that in its 2024-O counterpart.

4.3 Perforation Resistance of the FMLs

The impact response of the FMLs was investigated by determining the perforation threshold of each system investigated during the course of this study. Here, test specimens were subjected to increasing impact energies until complete perforation of the target was achieved. Figure 14 compares the perforation resistances of the 2/1 FMLs based on the 2024-T3 and 2024-O alloys where the perforation energy is plotted against the thickness of composite in the FML sandwich. In these laminates, the thickness of the composite core sandwiched between the outer aluminium plies was increased from approximately 0.5 mm to 6.6 mm. In the figure, the two points on the y-axis correspond to laminates based on two aluminium plies bonded by a single layer of unreinforced polymer adhesive. Therefore, these points effectively correspond to laminates with a composite core thickness equal to zero. Clearly, the perforation resistance of both types of FML increases with increasing

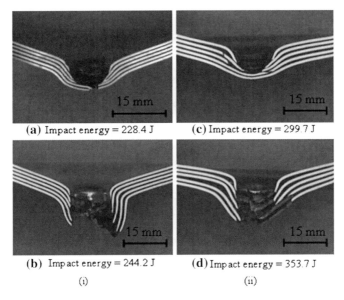

Fig. 12 Low magnification optical micrographs of polished sections of 5/4 FMLs subjected to various impact energies, (i) 2024-O 5/4 and (ii) 2024-T3 5/4

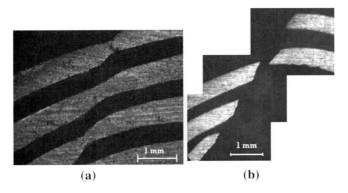

Fig. 13 Optical micrographs showing (a) thinning of the aluminium plies in a 2024-O laminate and (b) shear fracture of the aluminium plies in a 2024-T3 laminate

target thickness. For example, the perforation resistance of the 2024-T3 system with a 6.5 mm composite core was more that three times greater than that of the thinnest system. It is also evident that the 2024-T3 fibre-metal laminates outperform their 2024-O counterparts. The difference between the two laminate types is most pronounced in the thinner laminates, where the perforation resistance of the 2024-T3 FML is almost double that of its 2024-O counterparts. This difference

Fig. 14 The variation of the perforation energy with thickness of the composite core for the 2/1 FMLs. Included in the figure are the perforation thresholds of the plain composite. The solid lines represent the predictions of the perforation model

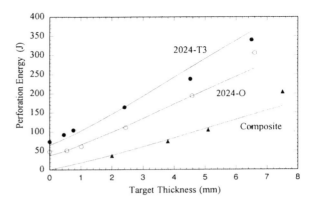

Table 3 Summary of the static linear elastic compression limit (σ_e) data for the families of systems examined here

Material System	Static linear elastic compression limit (σ_e) MPa
Plain composite	131
2024-O 2/1 laminates	192
2024-T3 2/1 laminates	247
2024-O multi-layer laminates	224
2024-T3 multi-layer laminates (0.8 mm plies)	338
2024-T3 multi-layer laminates (0.3 mm plies)	208

reduces as the laminate thickness is increased. It should be noted that the 2024-T3 aluminium is thicker than that of its 2024-O counterpart and this is likely complicate comparisons between the two systems. However, tensile tests have shown, however, that the higher strength 2024-T3 alloy absorbs considerably more energy up to fracture (i.e. a larger area under the stress–strain curve) than the 2024-O alloy and this appears to be translated into a superior energy absorbing capacity under impact loading. Included in Fig. 14 are perforation data resulting from tests on the plain composite material. Clearly, the impact resistances of both types of FML are superior to that offered by the composite material.

The solid lines in Fig. 14 represent the predictions of the Reid–Wen model. From the figure, it is clear that the model predicts the perforation resistances of the two types of FML as well as the plain composite with a high degree of success. Here, values of the static linear elastic compression limit (σ_e) of 131, 192 and 247 MPa were obtained for the plain composite, the 2024-O and the 2024-T3 systems respectively, Table 3. As expected the computed value of σ_e was higher for the FMLs than the plain composite and σ_e was higher in the stronger T3 FMLs than in the 2024-O laminates.

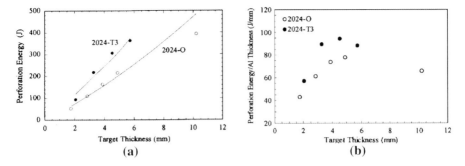

Fig. 15 a Comparison of the perforation resistances of the multi-layered 2024-O and 2024-T3 FMLs, b comparison of normalized perforation resistances of the multi-layered 2024-O and 2024-T3 FMLs

Figure 15a and b compares the perforation resistances of the multi-layered systems. Here, impact tests were undertaken on 2/1, 3/2, 4/3 and 5/4 laminates as well as on an supplementary 10/9 system based on the 2024-O alloy. As previously observed, the FMLs based on the stronger aluminium alloy out-perform their 2024-O counterparts, Figure 15a. Again, the properties of the aluminium alloy has the greater influence (in percentage terms) in the thinnest FMLs where the perforation resistance of the T3 FML was approximately double that associated with the corresponding 2024-O system. Figure 15b present the data shown in Figure 15a normalised by the thickness of the target. The data again support the conclusion that the T3 laminates (particularly at low target thickness) outperform their 2024-O counterparts. It is interesting to note that the data in Fig. 15b indicate that a laminate thickness of approximately 5 mm represent the optimum configuration. At values above this threshold, the normalised perforation energy begins to fall.

The solid lines in Fig. 15a correspond to the predictions of the Reid–Wen model discussed previously. The resulting values of σ_e are presented in Table 3 where it is again clear that the T3 system offers the higher value of σ_e. Once again, the perforation model predicts the impact resistances of the laminates with some success. Clearly, the model over-estimates the perforation resistance of the 10/9 system, this may be due to the use of an inappropriate value of σ_e for predicting the perforation resistance of this laminate. It is possible that the parameter, σ_e exhibit a thickness dependency, whereby it decreases with increasing target thickness. Further work is needed to fully investigate this.

Figure 16 compares the perforation resistances of T3 laminates based on thin (0.3 mm) and thick (0.8 mm) aluminium plies. The figure clearly shows that for a given target thickness, laminates with thicker aluminium plies offer a superior impact resistance to those based on thin plies. This results from the fact that, for a given target thickness, the volume fraction of aluminium in the FMLs based on 0.8 mm thick aluminium is greater than in the 0.3 mm FMLs. Given the fact that, for a given target thickness, the perforation resistance of the aluminium is superior to

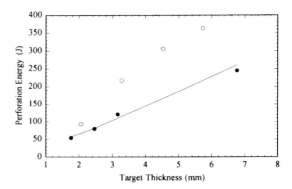

Fig. 16 Comparison of the perforation resistances of 2024-T3 laminates based on thin (0.3 mm) aluminium plies\it\text (*filled circles*) and thick (0.8 mm) aluminium plies \it\text(*open circles*)

that of the plain composite, it is not surprising that laminates based on thicker metal plies offer a superior impact resistance. If the data in Fig. 15 are re-plotted as a function of the thickness of aluminium within the composite, the two sets of data appear to collapse onto one line suggesting that ply thickness has a secondary effect on the perforation characteristics of these laminates. Included in Fig. 16 are the predictions for the perforation thresholds of the laminates based on 0.3 mm thick plies. Here, it is evident that once again the model successfully predicts the perforation resistance of these FMLs over the range of thicknesses considered. It is worth noting that the 6.7 mm thick system in this figure corresponds to a 10/9 laminate, a case where the model over-estimated the perforation threshold in the thicker 0.8 mm FMLs, Fig. 15a.

Figure 17 summarises the effect of stacking configuration on the perforation resistance of the T3 system. Here, the perforation resistances of the multi-layered systems are compared with those of the 2/1 laminates. The figure shows that for a given target thickness, the multi-layered FMLs outperform their 2/1 counterparts. As before, for a given target thickness, the multi-layered FMLs contain a greater amount of aluminium alloy and therefore offer a superior impact resistance.

In the final part of this study, the impact resistance of the SRPP FML system was compared to values determined from high velocity tests on similar materials. One of difficulties in comparing data from tests on different materials is that the target thicknesses and densities vary from system to system. Previous work to compare the impact resistance of fibre-metal laminates has involved the determination of the specific perforation resistance or what is sometimes referred to as an impact energy per unit weight [22]. This type of normalising procedure has been used to compare the ballistic impact resistance of fragment barriers and other types of arresting systems [23]. The perforation resistances of the various hybrid structures investigated in this research programme were therefore compared by determining the specific perforation energy (s.p.e.) of each laminate. Here, the s.p.e. values of the laminates were calculated by normalising the measured perforation energy by the areal density of the specimen.

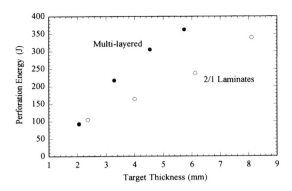

Fig. 17 Comparison of the perforation resistances of multi-layered FMLs and 2/1 FMLs based on 0.8 mm thick aluminium plies

Figure 18 presents the specific perforation energies of the FMLs investigated in this research study. Here, the FMLs are grouped according to the alloy type and stacking configuration (2/1 or multi-layer laminate. Included in the figure is the associated specific perforation energy of the aluminium alloy used for each family of laminates. The specific perforation energies of the 2/1 laminates based on the 2024-O alloy increase rapidly as the laminate thickness is increased. The s.p.e. of the thinnest 2/1 laminate is just above that of the plain aluminium alloy. However, as the thickness of the low density composite core increases, so does the specific perforation energy. The value corresponding to the thickest 2/1 laminate represented the highest value of s.p.e. recorded in this study. Interestingly, the s.p.e. values for the multi-layer 2024-O laminates were less impressive. Indeed, the s.p.e. of the 10/9 laminate was lower than that of the corresponding 5/4 system. Although the incorporation of increased amounts of aluminium increased the absolute perforation resistances of the FMLs, its relatively high density reduced their specific perforation resistances. The third grouping of laminates in Fig. 17 shows that the s.p.e. values of T3 2/1 laminates increases rapidly with increasing core thickness. It is interesting to note the performance of the thickest of these laminates was similar to that exhibited by the corresponding O laminates, a fact that emphasises the predominating contribution of the thick composite cores in these laminates. In contrast, the multi-layer T3 laminates out-perform their O counterparts, these being laminates in which the volume fraction of aluminium was high in all cases. Once again, increasing the thickness of these multi-layer systems eventually results in a reduction in the specific impact performance of these systems. The data in Fig. 17 show that the specific perforation energies of the thicker FMLs greatly exceed the values associated with the aluminium alloy.

Figure 19 compares the specific perforation energy of the 2024-T3 3/2 FML with previously-published data following impact tests on a range of woven thermoplastic-matrix composites (glass fibre reinforced polypropylene, glass/PEI, a glass fibre PA6,6, and a Kevlar PA 6,6 composite) and a thermosetting-matrix laminate (a woven glass fibre/phenolic) [16]. The superiority of the self-reinforcing

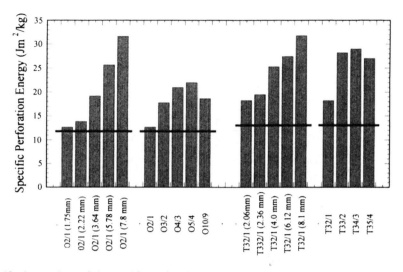

Fig. 18 Comparison of the specific perforation energies of the 2024-O and 2024-T3 (based on 0.8 mm thick aluminium plies). The dimensions in brackets correspond to the thicknesses of the 2/1 laminates

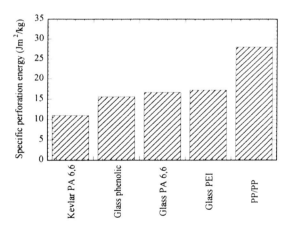

Fig. 19 Comparison of the specific perforation energy of the 2024-T3 3/2 FML with previously-published data

system examined here is clearly evident. It is interesting to note that the s.p.e. of this self-reinforced system is more than 250% higher than that measured on a comparable Kevlar-based FML. When the data are presented in term of perforation energy rather than s.p.e, the self-reinforced system continue to exhibit an excellent resistance to impact although the improvements relative to other systems are somewhat reduced due to the fact the SRPP system offers a low areal density. As previously stated, it is believed that the low modulus composite plies allow the metal layers to deform independently during the impact process facilitating energy absorption through membrane stretching and gross plastic deformation. The other laminates in Fig. 19 are based on stiffer fibres that may restrict this deformation process in the FMLs, thereby reducing their perforation resistance. The evidence in Fig. 19 suggests that FMLs based on combinations of thin layers of aluminium and SRPP offer significant potential for use in the design of impact resistant structures.

5 Conclusions

The high velocity impact performance of a range of fibre-metal laminates based on a self-reinforced polypropylene has been studied. It has been shown that these hybrid laminate systems offer potential for use in lightweight energy-absorbing structures. Impact tests have shown that multi-layer FMLs based on the stronger 2024-T3 alloy offer a superior perforation resistance to those based on a 2024-O alloy. The superior perforation resistance of the 2024-T3 system over the 2024-O is likely to be due to the superior energy-absorbing behaviour of the aluminium alloy in the former. A detail examination of the perforation zone indicated that significant thinning of the aluminium plies had occurred during the perforation process. It is postulated that the low modulus composite plies in the FMLs allow the aluminium layers to deform independently, absorbing significant energy through localised membrane stretching. In addition, the high strain to failure of the polypropylene fibres with the composites allow large amounts of energy to be absorbed in the failure process thereby enhancing the perforation resistance of these layered structures.

It was observed that the highest specific perforation energy was offered by a simple sandwich construction based on a thick composite core and thin outer aluminium plies. An examination of the failed laminates showed that membrane stretching, plastic deformation and shear fracture in the aluminium layers as well as plastic drawing, delamination and ductile tearing in the composite plies were the primary energy-absorbing mechanisms in these laminates. The perforation resistance of these polypropylene fibre-based FML structures can be predicted using a simple model previously used to predict the impact response of composite materials.

Acknowledgments The authors acknowledge the financial support of the Public Service Department (Malaysia) and Universiti Teknologi Malaysia. The authors are also grateful to Derek Riley of Propex Fabric for supplying the self-reinforced composite (Curv) and David Robinson and Professor Tony Johnson of Gluco Ltd., Leeds, UK for supplying the interlayer adhesive (Gluco), and to Mr. Peter Smith for his help in conducting the impact tests.

References

1. Peters, S.T. (ed.): Handbook of Composites. Chapman & Hall, London (1998)
2. Matthews, F.L., Rawlings, R.D.: Composites Structures and Materials. CRC Press, Boca Raton (FL) (1999)
3. Kelly, A. (ed.):Concise Encyclopaedia of Composite Materials (revised). Elsevier, Amsterdam (1994)
4. Prichard, J.C., Hogg, P.J.:The Role of Impact Damage in Post-impact Compression Testing. Composites **21**, 503–511 (1990)
5. Dorey, G., Bishop, S.M., Curtis, P.T.: On he impact performance of carbon–fibre laminates with epoxy and PEEK matrices. Compos. Sci. Technol. **23**, 221–237 (1985)
6. Zhou, G.: Prediction of Impact amage thresholds of glass-fiber-reinforced laminates. Compos. Struct. **31**, 185–193 (1995)
7. Cantwell, W.J., Curtis, P., Morton, J.: An assessment of the impact performance of CFRP reinforced with high strain carbon fibres. Compos. Sci. Technol. **25**, 133–148 (1986)
8. Thanomsilp, C., Hogg, P.J.: Penetration impact resistance of hybrid composites based on commingled yarn fabrics. Compost. Sci. Technol. **63**, 467–482 (2003)
9. Abrate, S.: Impact on laminated composite materials. Appl. Mech. Rev. **44**, 155–190 (1991)
10. Dorey, G., Sidey, S.G.R., Hutchings, J.: Properties of carbon fiber-Kevlar 49 fiber hybrid composites. Composites **9**, 25–32 (1978)
11. Krishnakumar, S.: The synthesis of metals and composites. Mater. Manuf. Process. **9**, 295–354 (1994)
12. Alderliesten, R.C.: Fatigure. In: Vlot, A., Gunnick. J.W. (eds.): Fibre–metal laminates: an introduction [Chap. 11]. Kluwer, Dordrecht (2001)
13. Reyes-Villanueva, G.: PhD thesis, University of Liverpool (2002)
14. Vlot, A.D., Kroon, E., La Rocca, G.: Impact response of fiber metal laminates. Key. Eng. Mater. **141–143**, 235–276 (1998)
15. Reyes-Villanueva, G., Cantwell, W.J.: The mechanical properties of fibre–metal laminates based on glass fibre reinforced Polypropylene. Compos. Sci. Technol. **60**, 1085–1094 (2000)
16. Cantwell, W.J., Wade, G., Guillen, G.F., Reyes-Villanueva, G., Jones, N., Compston, P.: The high velocity impact response of novel fiber–metal laminates. In: Proceedings of the ASME conference, New York (2001)
17. Reid, S.R., Wen. H.M.: Perforation of FRP laminates and sandwich panels subjected to missile impact. In: Reid, S.R., Zhou, G., (eds.): Impact behaviour of fibre-reinforced composite materials and structures, pp. 239–279. Woodhead, Cambridge (2000)
18. Morye, S.S., Hine, P.J., Duckett, R.A., Carr, D.J., Ward, I.M.: Modelling of the energy absorption by polymer composites upon ballistic impact. Compos. Sci. Technol. **60**, 2631–2642 (2000)
19. Sun, C.T., Potti, S.V.: High velocity impact and penetration of composite laminates. In: Miravete, V.A., (ed.): ICCM/9 Composite Behaviour, pp. 261–68. Woodhead, Cambridge (1993)
20. Bland, P.W., Dear. J.P.: Observation on the impact behaviour of carbon-fibre reinforced polymers for the qualitative validation of models. Composites: Part A . **32**, 1217–1227 (2001)
21. Available from: www.curvonline.com
22. Hagenbeek, M.: Impact properties. In: Vlot, A., Gunnick, J.W., (eds.): Fibre–metal laminates: an introduction [Chap. 27]. Kluwer, Dordrecht (2001)
23. Shockey, D.A., Erlich, D.C., Simmons, J.W.: Full-scale tests of lightweight fragment barriers on commercial aircraft, final report DOT/FAA/AR-99/71 (1999)

Index

A
Annealing, 94
Anti-symmetric, 45

B
Ballistic pendulum, 192
Blast resistance, 189
Blast response, 189
Blast, 2
Boundary conditions, 66
Bridging, 28

C
Carbon fiber/epoxy, 158
Carbon fiber-reinforced polymer (CFRP), 1
Carbon-epoxy, 145
Catalyst, 87
Cavitation, 18
Center-notched flexure (CNF), 144
Centroidal plane, 69
Characteristic length, 23
Coefficient of thermal expansion, 50
Cohesive element, 17
Cohesive zone, 2
Compactibilization, 87
Complex stress function, 6, 10
Compliance, 20, 112
Composites, 1
Compressive strength, 39
Constitutive model, 18
Contact force, 168
Crack closure, 5
Crack initiation, 153
Crack opening, 5
Crack propagation, 5
Crack tip, 13
Cross-ply, 139

D
Damage assessment, 135
Damage coefficient, 178
Damage dissipation energy, 51
Damage evolution, 136
Damage initiation, 2
Damage mechanism, 2
Damage parameters, 184
Damage propagation, 2
Deflection, 65
Delamination, 2
Dilatational energy density, 21
Discontinuous fiber, 88
Dissipation energy, 45
Distortional stress, 18
Double cantilever beam (DCB), 107
Double cantilever beam (DCB), 19
Drop-weight impact, 209
Drucker-Prager, 19
Ductile tearing, 231
Dynamic fracture toughness, 143
Dynamic response, 136
Dynamic stress intensity factor, 143

E
Edge delamination, 45
Effective crack length, 112
Elastodynamic, 150
End-notched flexure (ENF), 107
Energy release rate, 2

F
Face-sheet thickness, 190
Failure criteria, 157
Failure mechanism, 219
Failure models, 129
Failure modes, 129
Failure process zone, 20
Fatigue life, 2
Fatigue, 2
Fiber aspect ratio, 91
Fiber breakage, 134
Fiber bridging, 123
Fiber kinking, 18
Fiber metal laminate (FML), 219
Fiber orientation, 109
Fiber pull-out, 46
Fiber reinforced polymer (FRP), 108
Fiber volume fraction, 133
Fiber-matrix interface, 136
Fiber-metal laminates (FML), 1
Fiber-recruitment (FR), 85
Finite element, 6
Flexible core, 61
Flexural loading, 2
Flexural stiffness, 50
Fourier series, 70
Fracture mechanics, 17, 108
Fracture strain, 96
Fracture toughness, 2
Functionally graded (FG), 61
Glass fiber/epoxy, 189
Glass-epoxy, 145

H
Helmholtz free energy, 22
High-velocity impact, 137
Honeycomb core, 189
Hopkinson bar, 143
Hydrostatic pressure, 18
Hygroscopic property, 18
Hygrothermal, 108
Hygrothermomechanical, 108

I
Impact damage, 130
Impact energy, 160
Impact resistance, 139
Impact test, 133
Impact, 2
Impulse, 197
Incident energy, 226
Indentation, 61

In-plane compression, 158
Interface debonding, 6
Interfacial adhesion, 88
Interlaminar crack, 17
Interlaminar fracture, 107

L
Laminates, 2
Liquid crystalline polymer (LCP), 85

M
Matrix cracking, 2
Matrix yielding, 2
Maximum shear stress, 90
Mesh density, 50
Metal-composite system, 220
Microcracks, 46
Micromechanical model, 6
Micromechanics, 17
Mixed-mode bending (MMB), 110
Mixed-mode fracture, 109
Mixed-mode ratio, 107
Mode-I crack, 5
Mohr-Coulomb, 19
Moisture absorption, 107
Moisture environment, 108
Moisture expansion coefficient, 108
Moisture exposure, 108

N
Nondestructive technique, 135

P
Perfect bonding, 65
Perforation energy, 226
Perforation resistance, 221
Perforation threshold, 229
Phase transformation, 7
Plastic deformation, 85
Plastic flow, 18
Ply sequence, 54
Polycarbonate (PC), 85
Potential energy, 66
Power-law, 79
Projectile, 157

Q
Quadratic failure criterion, 46
Quasi-isotropic, 109

Index

R
R-curve, 107
Reinforcement, 130
Relative humidity, 108
Relaxation, 97
Residual stress, 50

S
Sandwich beam, 61
Sandwich structure, 1
Shape memory alloy (SMA), 5
Shear correction ratio, 70
Shear strength, 39
Shear-lag model, 89
Shear-sliding model, 93
Shock resistance, 190
Single-leg bending (SLB), 109
Slenderness, 98
Spalling, 27
Specific perforation energy, s.p.e., 237
Specific stiffness, 108
Stacking sequence, 109
Stress intensity factor, 5
Stress transfer ratio, 91
Stress transfer, 89
Stress-strain curve, 96
Swelling, 116

T
Tensile strength, 39
Thermal stability, 86
Thermoplastic, 86
Three-point bend, 45
Transition temperature, 9
Transversely isotropic, 145

U
Unidirectional composite, 45

V
Virtual crack closure technique (VCCT), 18
Volume fraction, 8

W
Work hardening, 99
Work of fracture, 47
Woven, 107